Creating Literacy Communities as Pathways to Student Success

Creating Literacy Communities as Pathways to Student Success offers a model for using literacy as a pathway for secondary students to explore fields from which they are often systematically excluded. In particular, this volume demonstrates how access for young Latina students to STEM related fields can be bolstered through engagement with mentors in writing and reading programs. Written for pre- and in-service teachers, as well as scholars across disciplines, this book aims to re-conceptualize the ways in which writing can best serve ethnically and linguistically diverse students, especially girls.

Jessica Singer Early is an associate professor of English Education at Arizona State University, USA.

Christina Saidy is an associate professor of English at Arizona State University, USA.

Creating Literacy Communities as Pathways to Student Success

Equity and Access for Latina Students

Jessica Singer Early and Christina Saidy

Routledge
Taylor & Francis Group
NEW YORK AND LONDON

First published 2019
by Routledge
52 Vanderbilt Avenue, New York, NY 10017

and by Routledge
2 Park Square, Milton Park, Abingdon, Oxon, OX14 4RN

Routledge is an imprint of the Taylor & Francis Group, an informa business

Library of Congress Cataloguing-in-Publication Data
A catalog record for this book has been requested

ISBN: 978-0-367-00108-7 (hbk)
ISBN: 978-0-429-44446-3 (ebk)

Typeset in Times New Roman
by Apex CoVantage, LLC

For Lucca and Evelyn

Contents

Acknowledgments

We would like to thank the National Writing Project for the financial support for Girls Writing Science. GWS was funded via support of the National Writing Project and National Science Foundation Intersections Grant (#1224161). We are especially thankful to Tanya Baker, director of National Programs for NWP, and Elyse Eidman-Aadahl, executive director of NWP, for helping us to envision and execute this project and the continued work of the Central Arizona Writing Project. The NWP has and continues to be academic home and support for our scholarship and teaching.

We are grateful to Metro Center Academy for giving access for this project and for being an ongoing partner in our research and writing project site. To the girls of GWS, we learned so much from and with you. Thank you for taking a risk in learning with us and showing what it means to examine the intersections of writing and science. To the teachers at GWS for inviting us into your classroom space, coordinating student schedules and permissions, and sharing your students with us.

We have a few personal acknowledgments:

Jessica: I would like to thank my husband, Jake Early, for his amazing support and sense of humor and Jax, Pippin, and Lucca for their joy, curiosity, and sweetness. I would also like to thank my parents, who encouraged me to be strong, smart, and fierce.

Christina: I would like to thank Mark Hannah for being an amazing partner. Mark, you are the first person I go to for bouncing around ideas and talking through research. Thank you for your emotional support and for being a wonderful partner in raising our kids. To Evelyn and Darren, who inspire me daily with their intellect, curiosity, and overwhelming love.

Introduction

Creating Access for Girls

> One thing that motivates me is the importance of being a role model as a woman scientist. I didn't have many examples of that growing up, and even in college and graduate school I wasn't sure if it was possible to be a scientist and also be a parent and a well-rounded person. I really love being a scientist, and a woman, and a parent. I want to show others, especially girls and women, that it is possible.
>
> —Katherine, archaeologist

We came to this project as women scholars, teachers, mothers, sisters, and daughters. We are also colleagues, collaborators, and friends. We are both English Education professors with offices down the hall from one another at a large public university in the southwest. We (Jessica and Christina) often begin our days together drinking coffee and bouncing ideas off of one another. As women in academia, these conversations frequently center on the challenges we face as women in the workplace and these shared experiences were fundamental as we developed the project we came to call Girls Writing Science (GWS). To help you see where we're coming from, we'll share a few of these experiences with you.

Jessica is a twin. Her twin brother, Matt, is now an attorney and an avid outdoorsman. Jessica and her brother grew up attending the same schools and were almost always in the same classes. Although they pursued different extra-curricular activities, through all the years of schooling, Jessica always thought school came more easily to Matt. She worked twice as hard to get grades, which were the same or not as strong as Matt's. She read all of the books assigned in English classes and he often did not. He wrote essays and passed tests about these texts based on class conversations. In class discussions, he was often called upon before her, even when her hand was raised first. Matt was good at articulating what he knew or at making up what he didn't know. Jessica grew up assuming Matt was smarter. Now, she still sees

Matt as smart, but has come to understand how people draw from different learning styles. Also, looking back, Jessica can see the ways in which Matt's strengths were often privileged over hers due to gender. Matt's intellect was often assumed and promoted in school. He could speak with authority when he raised his hand or when he just jumped in to share his ideas. He was called upon quickly and frequently, and he was listened to when he spoke. In the same classes, Jessica felt she had to be a hard worker, follow the rules, wait her turn, and practice what she said before saying it.

As a professor, Jessica has been continually surprised and frustrated by the obstacles she faces due to gender bias in academia. She is surprised because she grew up with a father who is a professor who never talked about or experienced tensions, mistreatment, or lack of opportunity based on gender. These conversations never happened because these experiences weren't happening to Jessica's dad in his daily work as a male professor. When Jessica began her career as an assistant professor, she was immediately told by more senior colleagues in her research area that she "should be a workhorse." Graduate students would come to her for help with articles and research and advising, but then choose male colleagues as their dissertation advisors. Early on, Jessica thought this must be because she was doing something wrong or because students didn't like her. Over time, she has realized that this is what happens to women in the workplace. Women are expected to be the workhorses. They are expected to be the readers, editors, cheerleaders, and guides, but rarely the leaders. And, they are expected to be nice.

In early childhood, Christina loved school. Her mom often tells complete strangers how Christina decided she wanted to be a teacher at the age of five. As a lover of learning, Christina often embraced subjects that came her way. She loved to read and write. But she also loved math and she was good at it. It was challenging and fun. In high school she loved the challenges of algebra and geometry, especially proofs. Her parents often thought it was peculiar that she loved math. Her mom would talk about how she had to repeat algebra because it was hard. In school Christina was in *regular* not honors math, which was largely reserved for students who had gone to more competitive feeder schools for elementary and middle school and had more advanced math instruction prior to high school. Despite her success and interest in math, she was always encouraged to pursue reading and writing as her strengths. Over time, Christina took the nudge. While she loves her current career, Christina has come to realize that early experiences allowed her to think of herself as a reader and writer more than as a mathematician. She also realizes these experiences were largely shaped by gender and class.

As a graduate student in Rhetoric and Composition at a large public university in the Midwest, Christina had the opportunity to teach a variety of writing classes. One of these classes was technical writing, a class required of undergraduate students in fields such as engineering, computer science, and the life and biological sciences. The first time Christina was assigned to teach the class, she examined her roster the week before classes started and was shocked to see 20 men and only 2 women on the roster. She talked with other instructors and graduate teaching assistants and found that while some had a few more women on the roster, the overwhelming majority of students enrolled in and requiring technical writing were men. This confirmed what Christina knew intellectually, that while women have had increased access to STEM fields in the last two decades, they are still woefully underrepresented in STEM at universities and especially in the workplace. Christina's experience teaching the class that semester further confirmed the disturbing gender issues at play. The women in the class, while clearly smart and confident in their subjects of study, were virtually silent in class. It was clear to Christina to survive in majors in which they were completely underrepresented, these young women had learned to keep their heads down and show their competence through their work. Christina's experience teaching tech writing illustrated the challenge many women face in pursuing STEM majors and careers. While these women clearly have the intellectual ability to do the work, they often must concede to a hidden set of gendered rules to accomplish their goals.

These stories are only a few of the ways in which we have and continue to face gender barriers in schooling and the workplace. With these experiences in mind, finding ways to increase equity and access for girls through literacy was at the forefront of our minds when we were granted a National Writing Project/National Science Foundation Intersections Grant (Baker & Carroll, 2016; NSF, 2012) to examine intersections between literacy and science. We were committed to using this project to support and create a literacy community to effectively provide access to underrepresented students, in this case, Latina girls in science. We merged the practice of literacy learning with an exploration of women in science to create GSW for a group of Latina adolescent girls at an urban high school.

This project values literacy as a point of access into disciplines too often made inaccessible for girls. It also provides a model for building communities of practice (Lave & Wenger, 1999; Wenger, 1998; 2006) to support young women as they encounter gender barriers in school, the workplace, and their communities. As we prepare students for the diversity of disciplines, concepts, topics, and ideas they will experience in the world, we want to find ways to use literacy to navigate, take ownership of, and use this information. Literacy has the ability to transform our everyday lives (Christensen, 2000; Early, 2006). Literacy practices help position us within or in connection to new discourse communities (Swales, 1988), allow us to imagine and plan for our future lives (Early & DeCosta, 2012; Saidy, 2013), and give us the tools to voice our ideas, interests, and aspirations.

Gaining Access for Girls

There is a pressing need to support ethnically and linguistically diverse girls as they envision, prepare for, and access their future lives (Bielby et al., 2014). More specifically, there is a great educational and societal need to support girls in learning to access, learn about, and succeed in science (Early & Flores, 2017; Saidy, 2017). Women do not enter science fields at the same rate as men for a variety of reasons, including not seeing these pathways as available to them, less accessibility to female science mentors, not knowing these pathways exist, or a perpetuated myth that girls "do not care" about STEM (Miller, Slawinski Blessing, & Schwartz, 2006; Scott & White, 2013; Sian & Callaghan, 2001). In recent years, there has been a dramatic increase in the number of women in college. For example, women now outnumber men in undergraduate and graduate programs, and more women are entering science fields than ever before (Burke & Mattis, 2007). However, men still far outnumber women in engineering, computer sciences, and physics. Similar gender preferences across STEM disciplines exist for graduate degrees, as women earned only 31 percent of physical science, 26 percent of mathematics/computer sciences, and 23 percent of engineering doctorates (Diekman et al., 2010; Hill, Corbett & St. Rose, 2010; Siann & Callaghan, 2001). While there is a pervasive myth that Latina women, and other women of color, are not interested in STEM fields (Ong, Wright, Espinosa, & Orfield, 2011), research disproves this and shows women of color who persist in STEM often do so because of supportive peer and community interactions and experiences (Espinosa, 2011).

A growing body of research and programs have emerged to describe girls' perceptions of their career options and to connect them more directly to science-related careers and pathways (Scott & Clark, 2013). Changing teachers' practices around areas such as standards-based teaching, hands-on learning (Harwell, 2000), and performance-based assessment (Freedman, 2002) has the potential to increase the attitudes and engagement of girls toward science (Parker & Rennie, 2002). For example, CompuGirls (Ashcraft, Eger, & Scott, 2017), a program offered at universities across multiple states, provides after-school, summer enrichment, and school-based programs for underserved secondary girls to explore STEM using innovative technologies like digital storytelling, computer programming and design, and robotics (for more information see: https://cgest.asu.edu/compugirls/programs). Programs like CompuGirls bring STEM opportunities to underserved girls with external grant funding and university support and their approach brings access to STEM with innovative and engaging STEM activities, some of which include digital and analog literacies. Even with these advancements in research and teaching, girls continue to have limited opportunities to learn about and engage with science compared to boys and, ethnically and linguistically diverse girls are less likely to pursue and enter science-related careers than their White peers (Scantlebury, Baker, Suga, Yoshida, & Uysal, 2007).

In this book, we offer a model of how teachers, literacy coaches, writing project sites, and university scholars may create literacy communities within or beyond their classrooms as a means of creating more equitable learning for ethnically and linguistically diverse adolescent students to imagine, investigate, and articulate their future selves (Lee, Husman, Scott, & Eggum-Wilkens, 2015). While the literacy community described in this book uses reading and writing to connect Latina girls to science-related careers, we want to emphasize the way this book can serve as a model for literacy teachers and researchers to create communities of practice for underserved students in connection to any discipline or content area with little to no funding or external support.

Writing Pathways

In the GWS project, we see writing as a way of giving secondary students ownership of their exploration into science and as a way of making their exploration public for peer, teacher, and outside audiences. In our work training and supporting English language arts teachers, we hear stories again and again of teachers feeling underprepared to teach writing and of feeling they need to focus their curriculum more on teaching reading, analyzing literature, and learning content. Many teachers feel they do not have the training or time for writing. Unfortunately, the challenges the teachers we work with face are not the exception. Recent research shows how secondary teachers across the US receive little to no professional development or pre- and in-service training on the teaching of writing (Bereiter & Scardamalia, 1987; Graham & Harris, 2005; Applebee, Langer, & Mullis, 1974/1986). Moreover, in today's schools, students across grade levels and at varying levels of writing abilities do not receive in-depth or sustained writing instruction on how to write regularly, expansively, or adeptly (Early & Saidy, 2014a/b). Adolescent writers mostly write to fill in worksheets, emulate formulas (e.g. five paragraph essay), or to prepare for and take standardized tests (Applebee & Langer, 2011; 2013).

Our concern is if we continue to leave the teaching of writing out of the secondary curriculum, then we leave out a key pathway from the content to the world beyond school. Through writing, students may communicate with different audiences and for varying purposes. They may use writing to imagine possible lives as individuals who have voice, opinion, understanding, and a stake in in the world. Writing is a fundamental tool in preparing students for their future careers in the sciences or otherwise. Atul Gawande (2007), a doctor and medical researcher, describes in *Better: A Surgeon's Notes on Performance*, how writing is part of his daily practice as a doctor and a crucial skill in supporting scientific thinking and scientific discourse:

> You should also not underestimate the power of the act of writing itself. I did not write until I became a doctor. But once I became a doctor, I found I needed to write. For all its complexity, medicine is more physically than intellectually taxing. . . . Writing lets you step back and think through a problem. Even the angriest rant forces the writer to achieve a degree of thoughtfulness. Most of all, by offering your reflections to an audience, even a small one, you make yourself part of a larger world.
> (p. 256)

Writing allows students to think about any discipline—art, math, science, history. Through writing, students can access the world beyond the classroom and see examples of how the content in their classes is used and practiced in the world. The more we prepare all students to communicate through writing, particularly in disciplines where women or ethnically and linguistically diverse students face barriers, the more doors will be open to them in their lives beyond high school. Teaching writing is a move toward creating more equitable and socially just classrooms and curriculum for all students. The act of teaching writing is the act of teaching students to learn and succeed in school and to access the world beyond school.

Writing to Learn

The past 40 years have brought increased awareness of the importance of learning to write and of students learning through writing across all disciplines (Britton, 1970/1992; Gunel, Hand, & Prain, 2007). Recently, the writing to learn movement in education has expanded our understanding of the ways writing can improve students' writing skills beyond the English/ language arts context to include all disciplines, including science. Writing to learn "assumes that writing is not only a way of showing what one has learned but is itself a mode of learning—that writing can be used as a tool for, as well as a test of, learning" (McLeod & Soven, 1992, p. 3). Research on writing to learn has shown the ways writing contributes to students' learning by helping them reflect and think critically about new information. When writing about subject matter, students increase their time on-task and increase the number of times they rehearse new information or concepts. Furthermore, writing serves as a self-monitoring strategy, allowing students to monitor their own comprehension of a topic or task, evaluate their own misconceptions, and change their ideas (Bangert-Drowns, Hurley, & Wilkinson, 2004; Klein, 2000).

Examples of research-based strategies for writing to learn, which are often used in science classes, include learning logs (Langer & Applebee, 1987; McIntosh & Draper, 2001) and quick writes (Gallagher, 2011; Kittle, 2008). Another writing to learn strategy is the use of microthemes (Bean, Drenk, & Lee, 1982), which are "mini-essays" on a focused topic. In a microtheme students might define a key term, summarize a research article, or respond to a debate. Like journals, microthemes can be evaluated quickly and holistically. These could be written on 3x5 notecards. Writing to learn can be used to start class discussions, summarize or ask questions about a lecture, respond to class readings, respond to peers, try out ideas to incorporate in a formal essay, and reflect on course content (Applebee & Langer, 2011; Atwell, 1990). Strategies employed in writing to learn are directed toward the goal of deepening understanding of content through writing. Newell (2006) explains, "When writing is construed as specialized genres that offer new ways of knowing and doing, the role of 'literate thinking' is expanded and deepened to include both the learning of content and the process of critical analysis" (p. 236). Writing to learn is fundamental as writing and content teachers work to teach writing in and across the curriculum.

Science and Literacy

Teachers across the disciplines are feeling increasing pressure to incorporate literacy into the content areas in meaningful ways. Science educators, in particular, have been tasked with using reading and writing about scientific concepts and material as a way to improve students' scientific understanding, learning, and communication about science. The recent release of the Next Generation Science Standards (National Academy of Sciences, 2012) also emphasizes the importance of embedding literacy practices and, more specifically, writing into the K-12 science curriculum as a way to prepare students for science-related fields of study and work. Furthermore, the National Research Council (NRC) has recently emphasized the importance of students conducting scientific investigations and explaining their findings in writing beginning in the elementary grades. Science literacy, as defined by the National Research Council (2012), includes the ability to read scientific texts and also the ability to record observations, to compose written arguments to support and defend results, and to use writing to communicate ideas and findings to real audiences and for real purposes.

The NGSS and the NRC align closely with the ongoing implementation of the Common Core State Standards in schools across the US and explicitly outline the connection between CCSS and literacy in science and technical subjects. The CCSS provide benchmarks for a number of reading and writing skills and applications students are expected to master and apply before graduating from high school to prepare for college and workplace demands. For example, beginning in kindergarten, the Common Core State Standards (CCSS) require students to compose informational and explanatory texts. Students are expected to do research, analyze and interpret information, form arguments, and convey their findings appropriately in writing, with evidence to support their claims. Additionally, with the CCSS (2011), there is a strong emphasis on literacy in science and social studies beginning in grade 6. There are wonderful teaching resources and materials to support science educators in bringing more innovative literacy strategies to the science classroom (Altieri, 2016; Elliot, Jaxson, & Salter, 2016; Saul, Reardon, Pearce, Dieckman, & Neutze, 2002). While the GSW project focuses on using literacy to access scientific careers and pathways, we demonstrate a model and case study of a literacy community which could be used to access any discipline and we hope will be taken up across the curriculum. The possibilities for creating literacy communities as points of access are endless.

A Literacy Community of Practice

In developing GWS, we sought to build a community of learners and practitioners engaged in inquiry and literacy activities to increase access to science fields and careers for young women, particularly first-generation college bound young women of color. In building this literacy community, we drew on Wenger's theory of communities of practice (Wenger, McDermott & Snyder, 2002; Wenger, 2006). According to Wenger,

> Communities of practice are formed by people who engage in a process of collective learning in a shared domain of human endeavor. . . . In a nutshell: Communities of practice are groups of people who share a concern or a passion for something they do and learn how to do it better as they interact regularly.
>
> (2006)

Communities of practice can be formal or informal groups, yet all share three common characteristics: 1) the domain, 2) the community, and 3) the practice. Domain is a shared domain of interest, commitment to that domain, and shared competence that distinguishes it from other people. Community includes joint activities and discussion, sharing of information, and relationships that enable community members to learn from one another. Practice refers to "a shared repertoire of resources: experiences, stories, tools, ways of addressing recurring problems—in short a shared practice" (Wenger, 2006).

In education, specifically in schools, communities of practice develop and are cultivated in a variety of ways. Much of the literature on communities of practice in schools focuses on the ways that teachers create communities of practice starting in pre-service programs (Sim, 2006) and in professional practice via professional learning communities (PLCs) (Vescio, Ross, & Adams, 2008) or teacher induction groups (Cox, 2004). For teachers of writing across the disciples, the National Writing Project serves as a community of practice (Whitney, 2008). The literature on communities of practice for students focuses less on in-school communities and more on the ways students experience communities of practice in school-community partnerships with an emphasis on civic engagement (Mitra, 2008). However, students often experience informal and formal communities of practice in school. Student clubs often develop as communities of practice. For example, a club such as a school robotics is a community of practice with a domain (robotics), community (engineering-minded students), and a practice that includes building, sharing, and competing. Students participate in multiple and diverse communities of practice—both sanctioned and unsanctioned—simultaneously throughout their schooling experience. Furthermore, communities of practice reach beyond the school walls linking students to other groups and communities outside of school that are linked by domain, community, and practice.

Communities of practice served as the overarching framework for this work. We also draw from complementary and intersecting theories, which we detail in future chapters. These theories include sociocultural theory, feminist theory, and genre theory.

Our Purpose

Building off of these theories, much has been written about the need for teachers, researchers, and schools to foster literacy opportunities for students to build bridges between classrooms and the worlds they will one day enter through higher education, the workplace, and the community (Purcell-Gates, Duke, & Martineau, 2007; Gallagher, 2011; Early DeCosta, 2012). There is also a growing body of research examining the benefit of exposing ethnically and linguistically diverse girls to hands-on and interactive science learning (Freedman, 2002). This project fills a gap in current research by offering a concrete and replicable model of how researchers and teachers may work toward increased equity and access in education by creating literacy communities of practice using reading and writing as a way for Latina students to explore their future selves in connection to a specific content area or discipline. Our guiding inquiry focused on examining the ways the Girls Writing Science community worked to invite students to dream, to ask, to network, and to envision their future selves in connection to science using literacy.

In presenting this model of a literacy community as a pathway for dreaming, asking, networking, and envisioning, we ask a series of interrelated, motivating questions: 1) What does it mean to create a literacy community in which students use reading and writing to connect to a discipline and envision their future lives in connection to that discipline? 2) How was this literacy community designed and enacted within a school space, and what might this look like in informal learning spaces? 3) How does this kind of literacy community form networks, utilize mentors, and build upon participants' interests and future goals? And, 4) In what ways may this project serve as a model for future school or out of school communities of practice collaborations to provide access for underrepresented student populations through literacy?

A Lens Into This Book

This book is for pre- and in-service teachers, informal and formal educators, and scholars working across disciplines who are interested in finding ways to provide opportunities for youth to use literacy to imagine, learn about, discover, engage with, and work toward their future lives. This book is also for National Writing Project teachers, literacy coaches, school leaders, researchers, and all those involved in education who want to see a greater emphasis on creating opportunities for students to engage in literacy communities where they take part in writing and reading as an avenue of engagement with mentors, information, and possibilities for their future lives. It is also for educators who want to promote equity and access for underserved or underrepresented students, with a particular attention to girls.

Chapter 1, "Creating a Literacy Community," describes the political, educational, school, and classroom context. We describe the students, classroom teachers, and ourselves in connection to this study. This chapter provides a logistical overview of what we did to make this project possible.

Chapter 2, "Learning to Dream," describes the first weeks of the project and the ways we invited the groups of girls to use writing to work through the process of dreaming their future selves in relation to science. Students drew representations of scientists, wrote their reasons for participating in the GSW project and wrote about success and hardship they had experienced in connection to science.

Chapter 3, "Learning to Ask," describes how the activities in GWS encouraged girls to make connections to historical women in science, express their interests and needs through informal writing, and create interview questions to ask for what they need from women currently working in scientific fields. This chapter employs a feminist lens to invite students to ask for what they need and advocate for their interests in supportive communities of women.

In Chapter 4, "Learning to Network," we share the ways we created opportunities for the girls to make connections to women in science through writing, interviewing, and technology. Building on the concept of feminist rhetorical ecologies, this chapter shows the ways that networks are built and sustained. Through this networking, the girls were able to begin to imagine their lives in science.

Chapter 5, "Learning to Envision," details the way we asked the girls to reflect on what they had learned through the act of interviewing women in science. In this chapter we share the girls' final interview profiles and the responses the women in science wrote to the girls who interviewed them. We also share final reflections on the interview writing project where the girls talk about what they learned and how they were impacted by the work we did together as a community.

In Chapter 6, we share final reflections and writing from the girls and their final drawings and imaginings of what it means to be a scientist. We conclude with a call to teachers, scholars, and informal educators to find ways to use this model to form literacy communities as pathways for students.

The model described in this book is intended to support educators interested in designing and implementing literacy communities of their own for students they know and work with who could benefit from engagement in a literacy community where the focus of the group is imagining, moving toward, and engaging with possibilities for their future lives. While in this book we explore a Girls Writing Science project for Latina girls, we believe this is a model that can be adapted and extended in diverse contexts for different populations. Above all, this book is a call to experienced educators to use literacy as a set practices to provide access and equity for students to ideas, content, life pathways, and opportunities. It is an invitation to secondary teachers—faced with pressure to demonstrate student success on high-stakes tests and to follow ever- changing state and national policy mandates—to think of reading and writing as vital tools for all students to prepare for their future lives and to have access to the world of possibility awaiting them beyond school walls. This model works to re-conceptualize the ways literacy can best serve ethnically and linguistically diverse students, especially girls.

1 Creating a Literacy Community

> I learned a lot from this project. I learned that being a girl in this field is a little difficult, but it's a challenge I'm willing to take.
>
> —Clara, tenth grade

When we received the NWP/NSF Intersections grant to develop a project on the intersections of science and literacies, we were first invited to a planning meeting in Colorado with other NWP site leaders from writing projects who received funding. We traveled to this meeting with our commitment to increased gender equity and inclusion at the forefront of our minds. The NWP invited Anne Richardson, director of Educational Partnerships at the Exploratorium in San Francisco, California to share her research on her educational program and professional development work to train informal science educators to be "field trip explainers" at the Exploratorium. Field trip explainers act as guides, teachers, listeners, and observers for individuals visiting the Exploratorium school, camp, or community program field trips. Richardson shared how the informal educators entered her Field Trip Explainer Program with perceptions of what counts as science learning, which had been influenced strongly by experiences learning science in formal school settings. They had an understanding of science learning as a rigid process reserved for a select group of people. Through participation in the Field Trip Explainer community of practice, Richardson worked with the informal science educators to develop an understanding of science learning as accessible, experiential, and prevalent (Richardson, 2012).

In her description of her research and program, Richardson described how she drew from Wenger's theory of communities of practice (2006) to create a community of practice among the informal science educators at the Exploratorium. Richardson, building a community of practice, helped bring about fundamental change in the way the informal science educators viewed their roles within the science center and, in turn, they shifted their approach to teaching science.

We were inspired to create a girls' science literacy community to provide access to STEM fields and shared this idea with our partner from the local science center, an informal science educator, woman, and mother to boys. Surprisingly, she saw the idea as exclusionary and unfair since it did not include boys. This experience of clashing or disconnecting lenses reminded us of how gender barriers also include the limits women put on other women. Because of our experiences as women and our knowledge about ways women are underrepresented in STEM fields, we did not see the project as exclusionary. However, we were confronted with a challenge often faced in collaborative work across institutions and disciplines. We needed to find a way to work together successfully even coming from diverse perspectives. We did not push our idea for a girls' science literacy project and, instead, focused on a different project in the first year of the grant. However, in the second year of the project, our partner at the science center moved on to a new job, and we found ourselves without a partner for a grant requiring a partnership. With the encouragement of NWP, we returned to our original idea to create a Girls Writing Science Project.

We wanted to create a literacy community of practice within a formal school setting to provide access for underrepresented high school girls to STEM careers. Our hope was to create a project that could easily be replicated by teachers, scholars, literacy coaches, or volunteers without the grant funding or external support we were benefiting from. At the heart of this project was the creation of a literacy community of practice with three central tenets:

1. This was a feminist project. It was a project by women, for young women, to connect to other successful women.
2. This project was about equity access. Often expert reading and writing practices are relegated to honors Advanced Placement or International Baccalaureate students, students who are perceived as "college bound" or students from a particular social class. This project was created for and committed to students who did not fit any of these categories. We view rigorous and real-world literacy practices as fundamentally accessible to all students regardless of the categories in which they fit. It is through reading and writing that students can gain access to the unfamiliar and to imagine what is possible in their future lives.
3. This project was about creating a literacy community. We do not see reading or writing as belonging to disciplines, professions, age groups, or standardized formats. Rather, we see literacy as a set of fluid and ever-evolving practices. This project focused on the way literacy can be used across age groups, disciplines, genres, and locations.

Setting and Context

Political and Educational Backdrop

Jessica's and Christina's view of literacy research and teaching is grounded in the belief that these acts take place in connection to the larger social, political, and historical contexts with which they occur (Early & Flores, 2017). At the time this project took place, our state stood at the epicenter of national immigration politics, which provide an important backdrop for our work in multiple ways. In the years prior to the year of our study, the governor had put forward an unprecedented anti-illegal immigration measure granting police the permission to stop, question, detain, or arrest based on reasonable suspicion of illegal immigrant status. In the years following, this highly contested legislation sparked a national debate about race, racism, citizenship, immigration, and civil rights. Undocumented immigrants living in the urban area became the target of immigration sweeps by federal and county enforcement officers, which resulted in regular police surveillance of predominantly Hispanic communities, including the city center and neighborhoods surrounding the school where this study took place. Leading these sweeps was the infamous and incendiary county sheriff known for his highly politicized and visible stance against undocumented immigrants. Because of this political situation, the school, along with schools throughout the state, experienced high numbers of absences and withdrawals as immigrant parents feared imprisonment or deportation for themselves or their children. Students and their parents are not legally required to disclose their immigration status; however, the culture of fear and uncertainty permeated into school contexts, particularly in schools with high numbers of Latinx students, like the one where we conducted this study.

Along with the highly charged political context, it is useful to understand the educational terrain surrounding this work. The state where this work took place falls close to last in the country for per-pupil funding, teacher salaries, and teacher retention. Lack of state educational funding trickles down to impact student preparedness, achievement, and graduation rates (US Census Data, 2017). For example, in our state's largest urban center where this study took place, there is strong evidence that nearly half of ethnically and linguistically diverse students do not meet the writing requirement of the state mandated high school competency exams. In response, urban districts in the state are desperate to find well-prepared, committed new teachers to educate their children to acquire critical writing practices (Guarino, Santibañez, & Daley, 2006).

The state has a large number of charter schools due to favorable charter school and open enrollment legislation dating back to 1994. Because of the state's support for school choice and charter schools, many parents have freedom to select schools for their children far beyond their neighborhoods. The majority of citizens in the state live within a very short drive from a charter school, and access to these schools is far greater than in many states across the country (Maranto, Milliman, Hess, & Gresham, 2008). When parents send their children to charter schools, the public schools lose the per-pupil funding for these students. While school choice, in many parts of the state, has led to more competition and variety in school options, it has also led to an exodus of middle- and upper-class White students from local neighborhood public schools, making these schools uncharacteristically racially homogenous (Nevarez & Wyloge, 2016). This is the case for the school, Metro Center Academy, where this work was situated.

Metro Center Academy

Metro Center Academy, a free-public charter school, located in the heart of a major metropolitan city in the southwest, serves students from kindergarten through twelfth grade (names of the school, students, and classroom teachers are pseudonyms). The school sits on the corner of a busy street in the heart of the city's downtown. It is surrounded by a high, chain-link fence and parking lots. A local university, in partnership with the city's high school district, sponsors the school's charter. The university chose to renovate an older, unused school building that sits on a busy thoroughfare in the middle of the city's downtown. The university is connected to Metro Center Academy in a variety of ways. Most of the teaching staff is made up of university graduates from the teacher training school or the university supported Teach for America Program, which is unsurprising since the university trains the majority of teachers in the metro area. At the time of the study, the university supported the salary of a professor in residence focused on literacy research and teacher training. The university, with permission, also uses the school as a test site for university sponsored grants, research, author visits, and sports and entertainment events.

The school is located across from the city's sports, theater, and art facilities. While there is little green space on the outside, the inside of the school is lined with student work, event posters, and open doors to classrooms. The sound of teachers teaching and students learning flows from classrooms. When you walk in the school, students or administrators often greet you, hold the doors open, and ask if you need help finding your way. We chose this school site because two women teachers from our first year of the grant project work there. One is an ELA teacher and the other a science teacher. We also have a strong relationship with the administration and a commitment to serving the student body.

The school community is largely Hispanic and made up of working class and low-income students. The student body comprises 1,100 students. The school is 68 percent Hispanic, 15 percent White, 14 percent African American, 2 percent Native American, 1 percent Asian, and 1 percent two races. Metro Center Academy receives Title I funds with over 74 percent of the student body qualifying for free or reduced lunch (NCES, 2013–14). Sixty-nine percent of the students come from homes where Spanish or another language, other than English, is spoken.

Through professional development, teacher training, and National Writing Project programs, we have formed a bidirectional relationship with Metro Center Academy and, more specifically, its English department. Using recommendations from our ELA and Science teacher contacts at the school, as well as an open invitation letter to parents, we initially recruited twenty-two girls in ninth, tenth, and eleventh grades. The teacher recommendation process helped in recruiting girls the teachers thought would benefit from a GWS project and the letter home to parents helped ensure family and student interest and investment (see Figure 1.1 for an example of a recruitment letter). There were no specific GPA or grade requirements to participate.

Unlike many schools in larger districts, Metro Center Academy does not have a strict curriculum for teachers to follow. While the school draws from the Cambridge Curriculum, teachers have great latitude in designing learning opportunities for their students and in working across the curriculum.

Recruitment Letter:

A Science Writing Workshop for High School Girls

Dear Metro Center Academy Student:

We are professors in the Department of English at the local university. We are inviting your participation in an exciting writing workshop and research study for 7th–12th grade girls. This workshop will take place once a week for six weeks during your Wednesday Capstone class time at Metro Center Academy from April 8th–May 13th. We are creating a literacy community for 9th–12th grade students to invite high school girls to use reading and writing to think about, imagine, and plan for future studies and possible careers in the sciences. Your participation will involve writing, reading, drawing, discussing, and goal setting.

Your participation in this project and study is completely voluntary. If you choose not to participate or to withdraw from the study at any time, there will be no penalty. For example, it will not affect your grade in any of your classes. For those individuals who are under 18, parental consent will also be required.

We look forward to hearing from you, or your teacher, if you are interested in participating.

Figure 1.1 Recruitment Letter

Metro Center also has a Capstone period built into the schedule. During this period students can be released to other teachers for tutoring or participate in club activities. The expansiveness of the curriculum and the Capstone period afforded us the opportunity to conduct GWS during the school day.

The Girls

Out of the 22 students who began, 15 girls completed this project. Participation was voluntary. Some students chose not to continue after the first few days, others were drawn away for assessments and testing, and some were absent due to illness. Of the fifteen participants, twelve identified as Hispanic/Latina, two as White, and one as Filipina. Thirteen of the fifteen girls spoke a second language, and seven reported speaking Spanish only at home. All of the students planned to attend college and thirteen were first-generation college bound. The girls reported GPAs ranging from 2.38 to 4.25 and the group exhibited a range of academic strengths (see Table 1.1). The girls were all granted release by their teachers from their normally scheduled Capstone courses, a combination of study hall, school spirit activities, and character education, to attend this classroom-based workshop once a week for seven weeks for an hour to two hours per session. The students who participated in the workshop came from an array of backgrounds and drew from a variety of experiences for inspiration. Some were soccer or badminton players, others were dancers. Some students loved hip-hop music, fashion, mother-daughter bonding time, and extra-curricular activities. Other students had part-time jobs after school and on the weekends or cared for younger siblings at home.

Metro Center is a busy place. The tone of the school is ever changing, and students are used to this. The girls arrived at our first meeting late because they were taking yearbook pictures in the library and some were finishing testing, and others were coming from a club. While we immediately felt a bit jumbled by all of the movement and activity, the girls easily transitioned in and out of the classroom and back again. Sometimes they were late because they had prior commitments or had not been told we were meeting, but then arrived and were set to go. On the first day, the girls were quiet and uncertain. Yet, they seemed eager and interested. Many of the girls did not know each other well because they were across grade levels. There were a few girls who sat with friends.

There was Mariana, a tenth-grade student who came to GWS excited to learn. Mariana claimed she loved everything about science. In fact, she loved science so much that she could not decide on a singular scientific field that interested her. Mariana always asked questions, contributed to discussion, and completed excellent work. Mariana had participated in a summer program for girls interested in science at the local university and was regularly

Table 1.1 Girls Writing Science Participant Information

Pseudonym	*Age*	*Grade*	*Ethnicity*	*First Language*	*Second Language*	*Language Spoken at Home*	*GPA (Self-Reported)*	*Parents Education Level*
Amelia	15	9th	White	English	Spanish	Spanish	N/A	Mother: Did not graduate high school Father: High school
Marissa	15	9th	Mexican American	Spanish	English	Spanish	2.5	Neither graduated high school
Carla	16	10th	Hispanic	English	English	English	3.0	Neither graduated high school
Anita	15	9th	Mexican/ Puerto Rican	English	Spanish	English	2.6	Mother: 4-year college Father: Graduated high school
Mariana	15	10th	Latina	English	Spanish	English	2.4	Mother: Did not graduate high school Father: High school
Alejandra	14	9th	Mexican/ African American	English	None	English	3.4	Mother: Did not graduate high school Father: High school
Ana	15	9th	Mexican/ African American	English	None	English	2.8	Neither graduated high school
Clara	14	9th	Hispanic	Spanish	Spanish	English	2.7	Neither graduated high school
Leah	15	9th	Mexican American	Spanish	English	Spanish	2.3	Neither graduated high school
Josephina	15	9th	White	English	Spanish	Spanish	3.6	Mother: Did not graduate high school Father: Graduated high school
Miranda	15	10th	Filipina American	English	Yes	N/A	2.5–3.0	Neither graduated high school

Alicia	14	9th	Hispanic	English	Spanish	English	3.3	Mother: Graduated 4-year college Father: N/A
Carola	15	10th	Hispanic	Spanish	English	Spanish	3.8	Mother: Graduated high school Father: Did not graduate high school
Daniela	15	9th	Mexican American	English	None	English	2.7	Both graduated high school
Aja	16	10th	Hispanic	Spanish	English	Both (mostly Spanish)	3.6	Neither graduated high school
Tina	16	11th	N/A	Spanish	English	Spanish	3.2	Mother: Graduated high school Father: Did not graduate high school
Laura	15	10th	Hispanic	English	Spanish	Spanish/ English	3.9	Mother: Graduated 4-year college Father: Did not graduate high school
Miranda	15	9th	Mexican	Spanish	English	Spanish	3.0	Both graduated high school
Sara	14	9th	Hispanic	Spanish	English	English	3.9	Mother: Graduated high school Father: Did not graduate high school
Ariela	14	9th	Hispanic	Spanish	English	Spanish	N/A	Mother: Did not graduate high school Father: N/A
Estrella	14	9th	Hispanic	Spanish	English	Spanish	2.8	Mother: Graduated high school Father: Did not graduate high school
Biana	14	9th	Hispanic	English	Spanish	English	3.0	Both parents graduated 4-year college
Coco	16	10th	Hispanic	Spanish	English	Spanish	4.3	Mother: Graduated high school Father: Did not graduate high school
Mariana	15	10th	Hispanic	Spanish	English	Spanish/ English	N/A	Mother: Graduated high school Father: Did not graduate high school

encouraged by her parents to pursue a college major and then career in the sciences. Interestingly, Mariana's GPA was only 2.8. This was surprising to us because the quality of Mariana's work was excellent and her breadth of knowledge was expansive.

There was Aja, a quiet student who seemed shy when she first came to GWS. Aja was one of the girls who seemed sure of her focus from the start. She was interested in psychology. Aja sat with her friends in class and would chat until the opening activity, but she was always quick to attention when instruction began. Aja was clear that she hoped to learn as much as possible in GWS.

There was Miranda who was recommended for GWS because her science teacher was hoping it would spark an interest for the future coupled with community support. Miranda had recently suffered a personal tragedy with the death of a family member. Miranda was smart, made good observations, and asked good questions when she was on task. She was initially resistant in GWS. We had to gain her trust and be patient with her. We learned that she responded well to humor, and we used this as a way in. Miranda learned much in GWS and was an active participant.

Creating a Team

The Teachers

Two classroom teachers, an English language arts teacher (Hallie) and a high school science teacher (Clara), were generous in allowing access to their students and curriculum. We had worked closely with each of these teachers through our National Writing Project site. Each teacher had attended the Invitational Summer Institute two-week professional development program on the teaching of writing at our university site and had worked with us on a team during the first year of our Intersections grant. Each teacher chose to support the project through observational and organizational roles. Prior to the workshop, the ELA teacher assisted in communicating and establishing the workshop schedule. The teachers also helped recruit and communicate with students, other teachers, and the administration. They copied and stored materials, set up technology, and occasionally worked with students in small groups.

At the time of this project, Hallie was a third-year teacher at Metro Center Academy. She is a white woman and was in her 40s at the time this project took place. Hallie came to teaching and Metro Center after a 15-year career in business. In her three years at Metro Center, she developed a deep commitment to the goals of the school and to the students. She had also taught a variety of preparations and grade levels, including courses such as: ELA 9 and 10, creative writing, psychology, and capstone. Beyond teaching she has been the faculty sponsor for student government and other clubs. Because of the expansiveness of her roles and responsibilities at Metro Center Academy, Hallie knows everybody and clearly understands the roles of administrators, faculty, and staff. As such, Hallie took primary responsibility in this project for administrative tasks, such as distributing invitations to students and parents, gathering permissions, identifying available capstone dates for meetings, and communicating with teachers releasing students for GWS.

In her professional capacity, Hallie is deeply involved in her local writing project site. She has completed the Invitational Summer Institute (ISI), an Advanced ISI, and most of the professional development opportunities available to local teachers. Hallie has also attended the annual meeting of the National Writing Project and sees NWP as a professional home. Because of her connection to the local writing project site, Hallie learned about the Intersections project and applied to participate. In year one of the project, Hallie developed a public service announcement project and unit that she continues to employ in her classes.

At the time of GWS, Carmela was a first-year science teacher at Metro Center Academy. She is a Hispanic woman who speaks Spanish fluently. At the time of the study Carmela was in her 20s. Carmela's path to teaching was influenced by her work with the Intersections grant. In year one of the Intersections project, Carmela participated as an informal science educator from the partner site, which was a local science center. Throughout the first year of Intersections, Carmela came to know the writing project teacher participants and expressed her interest in teaching secondary school. At the end of year one when a science teacher position became available at Metro Center Academy, Hallie persuaded Carmela to apply for the job and for concurrent teacher certification program at the local university. Hallie and Carmela had developed a professional relationship in year one of the Intersections project that continued in year two as Hallie helped Carmela navigate her first-year teaching and the culture at Metro Center Academy. During GWS, Carmela primarily observed the workshops and helped Hallie with administrative tasks. However, the participants often turned to Carmela for science knowledge when they had questions. Carmela also helped the girls identify and connect with some women in science for the interview portion of the project.

Who We Are

Jessica and Christina collaborated on the design and implementation of the Girls Writing Science Project. We are both English Education professors and the director and co-director of a local site of the National Writing Project at our university. Jessica is a White middle-class woman with a PhD who speaks Spanish conversationally. She taught high school English language arts in urban school settings in Oregon and college composition at a large public university in California prior to becoming a professor. Christina is a Latina and White middle-class woman who speaks Spanish conversationally and understands Portuguese. She has a PhD and she taught middle school and high school language arts in urban school settings in California and college composition at a large public university in Indiana prior to becoming a professor. As research partners, we collaborated to create the overall curricular map for the project, to gain access to the school, to receive research permissions, and to collect data. Within the project, we divided components of the curriculum design and instruction based on our individual research interests and unique expertise.

Although we entered the school community as outsiders in terms of race, social class, education, and professional role, we were committed to teaching and researching this project as participant observers (Spradley, 2016) and we worked to take on an emic perspective through our dual roles as co-teachers and researchers (Erickson, 1985). We were not complete outsiders because of our prior work at the school and because many of the teachers at the school had gone through our writing project summer institute. Along with co-teaching, we co-created the curriculum, purchased supplies and snacks for the project, communicated with the Hallie and Clara and the school administrators. We helped set and clean up the classroom each week. We also sent weekly reflections, suggestions, and reminders to Hallie and Clara, checked in regarding curriculum and supplies, and participated in all of the workshop activities when we were not teaching. We received parental and student consents to conduct the study and to collect data prior to the project launch.

We came to this project as frequent collaborators, co-directors of our local NWP site, and friends. While we have many intersecting and complementary research and teaching interests, we also come to our work with individual lenses and distinct training. Jessica is trained as a social scientist with a PhD in Language, Literacy, and Composition and Christina is trained in the humanities with a PhD in Rhetoric and Composition. We are both writing researchers united in our commitment to creating literacy communities via social justice lenses.

Creating the Community of Practice

The community of practice formed within GWS contained three primary characteristics: 1) The domain in this case was a shared interest in women advancing in science via literacy. As we will share in the following chapters, the students, teachers, researchers, and interviewees all shared an interest in increasing access to scientific fields for underrepresented women. 2) The community included the shared reading, writing, and interviewing experience—the shared practices of literacy. 3) The practice in GWS included the shared literacy tools, stories of women in science (primary and secondary), and shared struggle for access and equity.

Data Collection

Throughout this project, we collected data from the following sources:

Demographic Questionnaire

This instrument included questions about age, ethnicity, schooling, language use, formal measures of academic achievement (i.e. grades, test scores, etc.), and future educational plans.

Opening and Closing Questionnaires

The opening questionnaire included three questions asking participants: 1) What science classes have you taken or are you currently taking? 2) Why did you choose to participate in this science writing group? 3) What would you like to know about becoming a scientist? The closing questionnaire included five open-ended questions asking participants: 1) Using your own words and ideas, please describe what you learned from this workshop on women and science and writing interviews? 2) In this workshop, we paired you with a woman working in a science field for you to interview. What was helpful about this process? 3) During this workshop you learned about different women in science through children's books, excerpts from books, and TED talks online. What was helpful about these resources? 4) As a final project, we asked you to write about a profile piece about the woman you interviewed. What did you learn about writing through this project? What was challenging? What was interesting? 5) What did you learn about yourself through this project? Do you envision your future life as a young woman exploring science? Please explain. In addition, both the opening and closing questionnaires included a prompt for additional questions and comments. Although there were certain limitations to the questionnaire and survey instruments in gauging students' goals and experiences in the project, these instruments served as a mechanism for understanding students' literacy practices, writing interests, current and future goals, and perceived growth related to the project.

Written Texts of Four Types

1) Texts written and produced for the Girls Writing Science curriculum (see Table 1.2), 2) written field notes about our teaching and observations, 3) teaching and observation memos completed at the end of meetings, and 4) email communication between one another, and the teaching team.

Table 1.2 Written Texts and Genre Categories

Workshop Day Produced	*Text Category*	*Genre*
Day 1	Opening Questionnaire	Questionnaire
Day 1	Self-Efficacy Survey	Survey
Day 1	Draw/Write What a Scientist Looks Like and Explain	Illustration and Reflection
Day 1	Sticky Note	Reflection
Day 2	Personal Experiences with Science	Personal Narrative
Day 2	TED Talk Notes and Reflection	Notes and Reflection
Day 2	First Thoughts Letter	Letter
Day 3	Gallery Walk Scavenger Hunt	Notes and Reflection
Day 3	Interview Checklist and Brainstorm for Women in Science Interviews	Checklist and Brainstorm
Day 4	Creating Interview Questions	Interview Protocol
Day 4	Writing Professional Email	Email Draft
Day 4	Interview Question Peer Workshop	Peer Review/Feedback
Day 5	TED Talk Reflection	Notes and Reflection
Day 5	Revise Interview Questions	Interview Protocol
Day 5	Reflection on Interview Choices	Reflection
Day 5	Final Email to Woman in Scientist	Professional Email
Day 5	Response to Model Interview of a Woman Scientist	Reflection and Goal Setting
Day 6	Post Interview Reflection	Reflection
Day 6	Peer Review	Feedback
Day 7	Draw/Write What a Scientist Looks Like and Explain	Illustration and Reflection
Day 7	Success in Science	Personal Narrative
Day 7	Self-Efficacy Survey	Survey
Day 7	Final Interview Profile	Interview Profile
Day 7	Closing Questionnaire	Questionnaire

Interviews

At the completion of the project, to obtain information about the girls' experiences and impressions of the project, Christina conducted brief (20-minute) semi-structured, recorded interviews with six students (Brenner, 2006; Spradley, 1979). The interviewees were selected based on their availability to meet after the GSW workshops, their willingness to be interviewed, and their regular attendance at all GSW meetings. Interview participants were asked 15 questions (see Figure 1.2), which included 11 questions about their experiences with the project and four demographic questions. Interviews were audio recorded and then transcribed. Interview responses provided additional information and details to already collected data sources.

Girls Writing Science Interview Protocol

1. What types of writing do you do mostly in school? Which of these do you enjoy most? Why?
2. What types of writing do you do outside of school? Which of these do you enjoy most? Why?
3. What are your writing strengths?
4. What are your writing weaknesses?
5. Before this workshop, what experience did you have with conducting interviews?
6. What did you learn about doing interviews from this workshop?
7. How do you think you'll use your interviewing skills beyond this workshop?
8. What about this workshop was interesting to you when you first heard about it?
9. What was interesting to you about science at the beginning of the workshop?
10. What is interesting to you about science now that you're done with the workshop?
11. Do you think of yourself as a scientist? Why or why not?

Demographic-Type Questions

12. Do you plan to go to college? Where?
13. Has anybody in your family gone to college?
14. What do your parents do for work?
15. What job do you hope to do when you are older?

Figure 1.2 Interview Protocol

Self-Efficacy Surveys

At the beginning and end of the study, the girls completed writing self-efficacy surveys. These surveys built on previous surveys of self-efficacy (Early & DeCosta-Smith, 2011) and invited the girls to rank on a Likert-scale from 1 to 10 their confidence in their ability to use writing to research, communicate, and inform both in and out of science settings. The self-efficacy surveys provided a quantitative measure of students' confidence in writing at the beginning and end of the project. This quantitative measure was used to complement our qualitative findings.

Creating a Curriculum

Out of our commitment to building a literacy community, each meeting of the Girls Writing Science Project involved explicit teaching, modeling, and practicing of writing to link the girls to women in science careers (Early, 2017). The overall curriculum involved four major facets. One was a gallery walk of women in science and nonfiction texts such as autobiographies, biographies, memoirs, children's books, recorded interviews, and TED talks of women scientists. The second and third parts included preparing for an interview with a woman scientist and involved goal setting, a First Thoughts Letter, interview protocols, and writing professional emails. In the third part, students identified, emailed, interviewed, and wrote up their findings. The final component included an opportunity for students to envision their future selves. For this, students drew a visual image of how they imagined a scientist pre-and post-workshop and then wrote a reflection on their interview experience and how it impacted their vision of their future selves (see Table 1.3). The writing curriculum design and instruction represented a continual process of checking in with and responding to students' needs, interests, and perspectives, and it also served as an avenue for students to begin to articulate their interests in ways they had not done before. Students began to think about, name, and examine their interests and future goals in relation to science.

We have organized the chapters to capture the themes that developed as we created this literacy community and provided access for Latina girls to science careers: dreaming, asking, networking, and envisioning.

Table 1.3 Curriculum Calendar Girls Science Writing Curriculum Calendar

Workshop Week	*Focus*
1	**Welcome** 1. Overview of the workshop 2. Draw a scientist (http://en.wikipedia.org/wiki/Draw-a-Scientist_Test) 3. Write what a scientist looks like and does 4. Writing workshop prompt—a time you felt success with science a time you felt shut out of science 5. Self-efficacy questionnaire 6. TED talk about science and the broadness of it—respond (www.ted.com/talks/asha_de_vos_why_you_should_care_about_whale_poo?language=en) 7. What questions do you have for her? What do you want to know? What do you want to know about her path as scientist? About her personally?
2	**Possible Selves and Strategy Boards and Interviews—Becoming a Woman Scientist** 1. Opening children's book (Title) 2. TED talk (6 minutes) (www.ted.com/talks/asha_de_vos_why_you_should_care_about_whale_poo?language=en) 3. TED talk worksheet—What questions do you have for her? What do you want to know? What do you want to know about her path as scientist? About her personally? 4. TED talk posters—write up responses 5. Introduction to Interview Assignment 6. First Thoughts Letter

Workshop Week	*Focus*
3	**Overcoming Barriers, Finding Sponsors, and Turning Points for Women** 1. Gallery walk of women in science—take notes and fill in Scavenger Hunt worksheet 2. If I could talk to any of these women in the room who would I talk to? 3. What is this telling me about what I want to know about science? Who might I want to interview? 4. TED talk: Lina Colucci 5. Reflection on TED talk—How are science and ballet related? What is something you are interested in that is related to science? 6. Interview protocol practice 7. Strategy Boards—Goal setting
4	**Interview Protocols** 1. Children's book 2. Tree scientist TED talk and reflection 3. Writing Interview Questions 4. Interview Question Peer Workshop 5. Writing Professional Emails
5	**Interview Protocol 2** 1. Children's book 2. Refine Interview Questions and send emails/make calls 3. Writing from our interviews—Bionic Woman write-up 4. How can I see myself in this interview? What is my possible life?
6	**Possible Selves** 1. Children's book 2. Writing your lead 3. Writing leads activity 4. Integrating quotes 5. Writing quotes activity 6. Making the turn—connecting yourself to a woman in science 7. Sharing drafts/revising
7	**Conclusion** 1. Review of the workshop 2. Draw a scientist 3. Write what a scientist looks like and does 4. Writing workshop prompt—a time you felt success with science a time you felt shut out of science 5. Self-efficacy questionnaire 6. Closing activity—sharing your profile and connecting to a woman in science 7. Reflection 8. Closing Certificates and Cake

2 Learning to Dream

> Today I remembered that science still interests me. Today I remembered that we can do whatever we put our mind to.
>
> —Clara, ninth grade

Standing at the front of the classroom holding *The Watcher*, Jessica introduces the children's book she is holding about Jane Goodall, a famous anthropologist and animal rights activist (Winter, 2011). She introduces the book by describing a memory of her mom taking her to hear Goodall speak at a local theater in Eugene, Oregon, when she was nine years old. Jessica shares how attending this talk deeply inspired her as a young girl. "After I heard Jane Goodall and saw images of the apes she worked with in Africa, I remember thinking, 'I want to do research when I grow up. I want to be engaged and helpful in the world in some way like Jane Goodall.'"

Twenty-two adolescent girls sat at their desks as Jessica read the story of this famous woman scientist on the first day of the GSW project (see Figure 2.1). The room fell silent. "'Jane, Jane, where are you?' "Jane can you hear me?' Everyone had been searching for hours and hours, looking for little Valerie Jane Goodall (p. 1)." Jane is full of curiosity even as a young child. She falls in love with animals, beginning with her own backyard chickens. As an adult, Goodall devoted her life to researching apes in Africa. As an anthropologist and activist, she advocated for the protection and preservation of animals through science. As Jessica finishes reading, she asks the group of adolescent girls if they have ever heard of Goodall before this story. No one responds. Then, Marissa, a ninth-grade student, shoots her hand up and blurts out, "I never knew there were people who studied monkeys for a job! I want to do that! How do I do that?"

An intention in creating and implementing the GWS project was to provide girls with opportunities to use literacy as a way to imagine and dream themselves in connection to science careers and pathways. Through this act of sharing her childhood inspiration, Jessica also modeled a story of herself being

Figure 2.1 Miranda listens intently as Jessica reads *The Watcher*
Source: Charlie Leight

touched by another woman's life pathway. By sharing memory and inspiration, the children's book about Goodall dreaming also became an example of a young girl dreaming herself into being. It is not just a story of a girl day dreaming or imaginary playing, but of a girl trying on and dreaming possibilities for herself in relation to the world. We use the term "Dreaming into Being" as a foundational and theoretical concept to ground this project. This term describes the enactment of a literacy community working to provide opportunities for youth through writing, reading, instruction and mentoring to imagine and dream themselves in connection to future pathways. In this project, we see literacy as an avenue for students to dig more deeply into the familiar as a way to gain access to the unknown. Within the GSW community, the girls used reading, writing, researching, and drawing, to learn about, to dream and envision, and to experience possibilities for themselves in relation to science, a field too often made inaccessible, intimidating, and unavailable to women.

Possible Selves and Sociocultural Theory

This project draws from two intersecting theories: sociocultural theory (Prior, 2006) and the theory of possible selves (Oyserman, Terry, & Bybee, 2002). We draw from sociocultural learning theory in which the specific cultural activities of language learning and writing are products of social interaction and are embedded in a larger cultural and institutional context (Vygotsky, 1978). From this perspective, the everyday worlds, social interactions, and relationships which students participate and engage are rich and valued sites of learning. A sociocultural lens also allows for an examination of adolescent girls engaged in a new kind of learning within the writing classroom, which is the social practice of forming and participating in a writing community as a means of imagining their future selves in connection to science.

Along with sociocultural learning theory, this study draws from Oyserman, Terry, and Bybee's (2002) concept of possible selves, which they define as "the future oriented component of the self-concept" (p. 313). Envisioning possible selves is the act of examining interests, goal setting, and situating self in the future in newly imagined ways with guidance and support. "Youth construct possible selves by synthesizing what they know about their traits and abilities and what they know of the skills needed to become various future selves" (p. 313). This study examines a secondary writing community which worked to tap into this notion of possible selves by helping Latina adolescent girls envision and think about their future identities, pathways, and interests in connection to science. We use Oyserman et al. (2002) concept of possible selves as an inspiration for what we call a "future lives writing curriculum," one that opens future possibilities for students to engage in planning for and envisioning their future lives or selves through writing (Early & DeCosta, 2012; Early & Hubbard, 2003).

The initial drawing and writing activities of the GSW project gave us a sense of the girls' understanding of themselves in relation to science. Their drawing and writing revealed the ways learning and identity are socially mediated processes and deeply rooted within the social contexts, contacts, and connections we are exposed to through school, family, and community. From this perspective, writing is always situated within a context, tied to specific purposes, and changes over time with practice and guidance (Bazerman et al., 2017; Dyson, 2003; Wardle & Roozen, 2012; Winn, 2011). This lens allows for an examination of the varied language and literacy practices at play in forming and enacting a literacy community (Early & Flores, 2017). The initial writing, drawing, and reading on the first day of the project, revealed the ways the girls' perceptions of themselves in relation to science were shaped by prior social and contextual forces (Bakhtin, 2010; Bazerman et al., 2017; Volosinov, 1973).

Using Writing to Dream

At the heart of this curriculum is a commitment to the power of literacy to build community, to learn, to communicate, and to open access to new worlds, spaces, and people. As writing researchers, we subscribe to a broad view of writing. Writing is not just putting words to paper. It is drawing (Heard, 1999,

2016), reflecting (Early & Saidy, 2014a/b), and using technology (Hicks & Turner, 2013) to generate ideas and respond to text and content. We believe writing can be taught and is not a gift of innate intelligence. It is a combination of many skills (Graham & Perin, 2007) acquired over time (Bazerman et al., 2017) with support, guidance, and practice (Brandt, 2001). As Charles Bazerman (2009) explained in his keynote address at the Conference on College Composition and Communication, "The Wonder of Writing,"

> Young people need to see writing in their worlds to imagine its possibilities; they need access to tools in order to learn to use them; they need nurturing, training, mentoring; they need endless hours of work, practice, and development.
>
> (p. 572)

Writing is a literacy skill for everyone. The more students are exposed to its diverse forms, purposes, and rewards, the more they may access different communities, disciplines, and lenses beyond their own. Writing provides a way for students to expand their worlds by engaging with new ideas, people, and perspectives in the world. Writing offers a way to dream.

This book draws on specific strategies for teaching writing in its many forms. In this chapter, we focus on ways to use writing to dream and to build community. To do this, we intentionally selected writing invitations throughout this project that drew on students' interests, lived expertise, and questions. This chapter describes the curricular choices we made early on to build community and set us up to do the work of moving outward beyond the classroom.

Dreaming Future Selves

In the first weeks of the project, students used writing to work through the process of dreaming their future selves (Oyserman et al., 2002) in relation to science. In the first meeting, we began by asking the girls to write their reasons for choosing to participate in the writing community. This writing revealed how some of the girls came to the project because they already envisioned their future lives in connection to science based on lived experiences and relationships. For example, Amelia wrote, "I want to interview an oncologist because I want to become a pediatric oncologist in the future. I have always found cancer so interesting. I have had cancer in my family." Another student, Clara, described her interest in interviewing a botanist. She had specific things she wanted to gain from her interview in terms of accessing information and gaining a better understanding of this career path. "I want to be a botanist in the future, so this interview can help me with my future job. Also, the woman I interview can help me understand what is required for a degree in botany."

Although some of the girls came to the project with clear direction and goals for their future lives in relation to science, the majority made it clear through the initial writing invitation that they had never dreamed their future lives in connection to science prior to this project. For these students, initial participation in this group and imagining their future pathways or interests in relation to science was part of a social process of connecting to teachers or with friends, more than a derivative of their own interests or future goals. For example, two of the girls expressed a strong interest in makeup and cosmetology because this is something they enjoy with their friends. They tried, in their goal setting writing, to connect this interest to science. Rosario wrote, "I would like to interview a scientist who does chemicals in makeup or who does makeup. I would like to interview her because it seems interesting and I feel like this person will be connected to me." The initial writing, for students like Rosario, revealed how dreaming in connection to science was new and challenging.

The writing also revealed how these students needed help honing a science-related interest/career path and narrowing overly broad visions. For example, Mariana wrote in an early reflection, "I'm fascinated in chemistry, physics, engineering, and sustainability." The writing also revealed a need for support to understand and define what counts as science and what it means to work in science. The writing curriculum design and instruction represented a continual process of checking in with and responding to students' needs, interests, and perspectives, and it also served as an avenue for students to begin to articulate their interests in ways they had not done before. Students began to think about, name, and examine their interests and future goals in relation to science.

Drawing to Dream

A girl wearing safety goggles and lab coat/cape, a bald man with poufy eyebrows wearing a lab coat, a woman with her hair tied tightly in a bun holding a glass test tube, are just some of the images the girls created when we invited them to draw images of what they thought a scientist looks like during the first workshop (see Figure 2.2). We came to this invitation as a way to gain a sense of what the girls already thought of as a scientist prior to the work of the GSW (Chambers, 1983; Finson, 2002). We set out blank paper, colored pens, and crayons at each table group. We also invited the girls to write in response to their drawing with the following question, "Now that you've drawn a scientist, take some time to write down what a scientist looks like and what a scientist does." Ten girls drew images of women scientists wearing lab coats and holding test tubes and beakers in lab settings. Mariana not only drew her woman scientist, but she labeled her drawing to explain each detail (see Figure 2.3). She included glasses, hair pulled up, nametag, pocket protector, lab coat, and beakers with chemicals. Four girls drew images of men in lab coats or suits. And, nine of the girls drew stick figures or people

Drawing a Scientist

Step 1: Think of what a scientist looks like and then take a few minutes and draw the scientist you imagine on this paper.

Step 2: Now that you've drawn a scientist, take some time and write down what a scientist looks like and what a scientist does.

Figure 2.2 Drawing a Scientist Invitation

covered up in gas masks. In response to the writing prompt, the girls shared what they brought to the GSW community in terms of their understanding of what scientists do and what "counts" as science. Alejandra captured her definition of a scientist using a bulleted list:

- Goggles and safety lab clothes.
- Concentrated and focused on their work.
- A scientist pretty much, depending on their field, investigates different subjects such as chemistry and biology.

Alejandra's definition included what scientists wear and how they pay attention to their subject of study. But, it also revealed how her main understanding of what counts as science came from her experience with science within a school context. We cannot assume this is the only science exposure to science

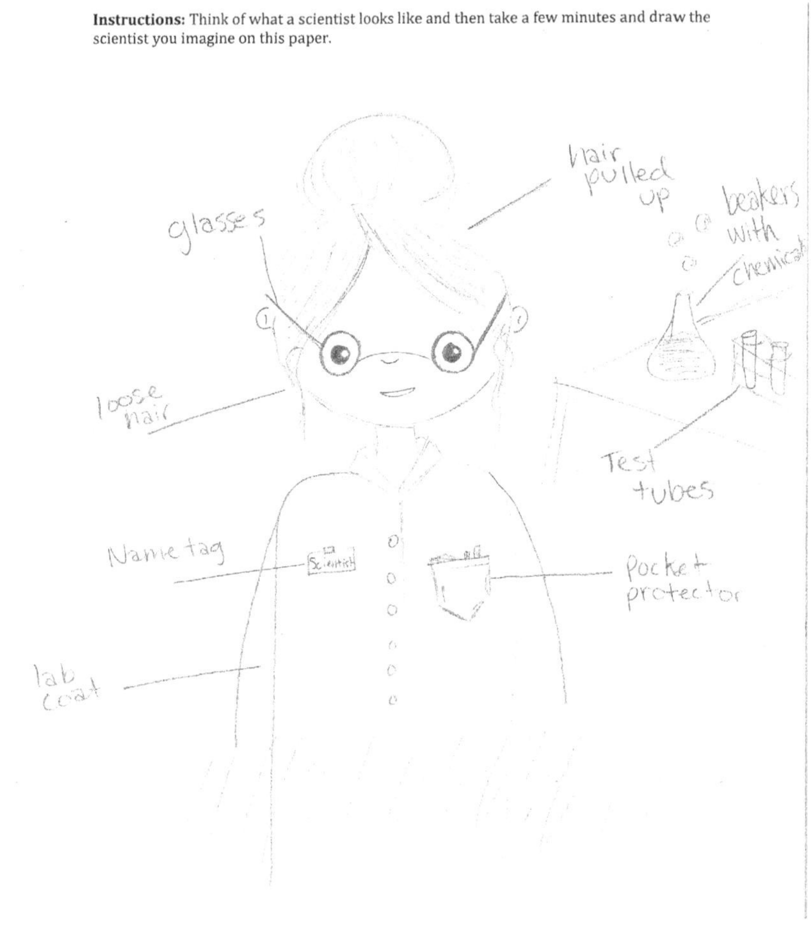

Figure 2.3 Mariana's drawing of a scientist

these students had had in their young lives; however, it is what came to mind in this early activity. Other students focused their drawings on what scientists wear and listed things such as lab coats, glasses, hair ties, and pocket protectors. Other students, like Leah, used the writing to describe the importance of scientific work and how scientists do a variety of things in their work. "Some scientists test substances, some look for a cure, other research. Yet, what they all have in common is that they try to evolve the human in a positive way." A number of the girls wrote about their understanding of scientists as intelligent, smart, important people, or even genius. Coco wrote, "A scientist is a genius with a white apron. He is always messing with all kinds of chemicals and asking a lot of questions." Only one of the girls wrote about the joy or pleasure she envisions from working in a science-related field. "She [the scientist] would be happy because she would enjoy her job."

In this initial writing activity, we were delighted to see so many of the girls envisioned women as scientists from the get-go and that they saw this work as important and substantive. We noticed the drawings were limited to lab coats and labs, or asking important questions, or being intelligent. These images reflect a schooled or public understanding of science. Generally, in the public sphere, we imagine scientists as lab workers in white coats. We could see through this brief window into students' perceptions of what it means to be a scientist, that they had an understandably narrow idea of what counted as science work and they also saw scientists as removed from their everyday lives. None of the girls drew or wrote about someone they knew or about a place they frequented or spent time in related to science. We saw, through this brief drawing and writing opening, how there was much room in terms of instruction, curriculum, and experience to expand the girls' understanding of what "counts" as science work and what scientists do.

Experiences in Science

In our first writing invitation to the girls, we asked them to write about their prior experiences with science. We did not want to make assumptions. It was also a way to start with personal experience, stories, and connections to science. We invited the girls to write in response to the following two questions:

1. Write about a time you felt shut out of science. This could be in or out of school.
2. Write about a time you felt success with science. This could be in or out of school.

These questions allowed the girls to share a range of experiences and to know that they could join the community with whatever experiences, interests, or apprehensions they carried.

Feeling Shut Out of Science

Through this writing invitation, we found the number one reason girls reported first feeling shut out of science was experiencing the feeling that it was "too hard" or "not understandable." Girls wrote about memories of failing science tests or missing science class and then "feeling lost." Leah wrote, "I missed two classes and I got assigned homework and I had no idea what to do. Not knowing what to do made me feel shut out of the topic and made me no longer want to try." For girls like Leah, she placed the misunderstanding or blame for not knowing what to do on herself for missing the class. She did not mention any opportunity or sense that she could receive help to get caught up with the material. Her understanding was that she either had to stay caught up herself or fall behind and stop trying and succeeding.

Girls also described an increasing lack of self-confidence and a growing sense that science was "not for them." For example, Coco wrote, "When I failed my Science 2 exam 3rd quarter, I felt shut out of science. I got a 60% on it and right after I saw my 60% and I wanted to stop science. I didn't think I could achieve in science. Science is a very hard subject for me."

Many of the girls described low grades or scores on exams as a self-fulfilling prophecy. They exhibited, through their writing, that the scores somehow reflected what they already knew or felt about themselves in science, and the ways it felt too challenging for them and out of reach. For example, Anita wrote, "I felt shut out of science when we had taken a quiz out of 10 points and I scored 3. That day made me feel so dumb. I just always knew science really wasn't for me."

Other girls, like Felicia, described how a lack of confidence related to science felt like an internal negative voice in her head that made her shut down. "I felt out of science in the beginning of this year because I didn't understand how to do an experiment. And, I started saying to myself I couldn't do it and that I couldn't understand what was going on." Five of the girls wrote about feeling shut out of science because of science exams that felt unclear or impossible. Amelia wrote, "I was completing a science exam and I did not understand the topic that we were working on. I had studied for the test, but I still did not understand it. This was one of my finals. I shut down science for a while after that." Only one of the girls wrote about the connection between not receiving help and not understanding or feeling shut out of science. "I feel shut out of science when I don't get the material and if someone doesn't help. Then I give up."

Many of the girls placed the blame or set the impetus for feeling disconnected from science on themselves, even when they consistently wrote about external factors such as tests, grades, lack of help or direction, or lack of understanding of what to do, as the problem. We thought about the ways the girl's responses to this first question were connected to recent research on girls' perceptions of intelligence, which develops at a young age. Schools throughout the country have recently adopted Dweck's growth mindset model (2006) to try to increase student achievement and efficacy. According to Dweck, a

> growth mindset is based on the belief that your basic qualities are things you can cultivate through your efforts, your strategies, and help from

> others. Although people may differ in every which way—in their initial talents and aptitudes, interests, or temperaments—everyone can change and grow through application and experience.
>
> (p. 7)

This model is particularly appropriate for encouraging students, especially women, to take on challenging tasks in STEM and beyond. However, our experiences with the girls' obstacles in science, remind us that gender is at play early on, even when a growth mindset is used. For example, a recent report by Bian, Leah, and Cimpian (2017) found between ages five and seven children start to believe brilliance is tied to gender and that girls are less likely to see their gender as brilliant. "Many children assimilate the idea that brilliance is a male quality at a young age. This stereotype begins to shape children's interests as soon as it is acquired and is thus likely to narrow the range of careers they will one day contemplate" (pp. 2–3). Bian et al. also found girls were willing to work hard, but not, necessarily, think of themselves as "really, really smart." In considering the responses of the girls in GWS, we realized both of these studies were relevant to our work. In a way, the girls felt shut out of science because they believed they lacked intelligence, or they doubted their intelligence even when scoring high on science exams. They saw these achievements as flukes or "lucky" rather than hard earned. Through these personal stories, we saw the need to both employ a growth mindset in our literacy community, but also to reinforce and support intelligence and interests when opening access to science through literacy.

Science Successes

As we read through the girls' written responses to the first writing invitation, it was striking to see how many of their successes were defined by school and by what other people thought of their science ability or how they performed on a test or a grade they received in a class. For example, Clara wrote about the astonishment she felt when a teacher recommended her for an Advanced Placement Science class, "My former teacher, Ms. A., told me I was going to be in the Advanced Placement Science class and I honestly didn't believe it at first, but it happened! My sisters were really proud of me and Ms. A. encouraged it." Estrella wrote about receiving a top grade in her science class and the pride she felt for this accomplishment: "When I had a 96% as my grade in science, I was the highest in my advanced science class. I made my science teacher so proud. Seeing her smile and her happiness for me made me feel amazing and proud." Seven students wrote about a shift that took place in their experience with science from feeling positive about it to disliking it. They wrote about earlier times in their lives when they "used to understand" or "did well" or "comprehended" science and then how something changed, and it became less accessible. For example, Coco wrote, "In the beginning of my sophomore year, science was very easy for me. I understood it very well. I passed most of my coursework and some of my exams. Then, toward my 3rd quarter, I started to not do so well." Rosario wrote, "A time I felt I succeeded in science was back in seventh grade. I would actually get good grades in science. I wouldn't need help from anyone." Rosario's response implies that she feels less success with science now because she needs more help and her grades are not as high. Her understanding of ability is tied to the idea that she should be able to do something challenging on her own without any guidance, that she should just innately know. Other students wrote about feeling successful with science when they suddenly figured something out or the material clicked, and they understood it clearly. For example, Alejandra wrote, "When we would look at the notes we take in school and I would end up getting what my teacher was saying and actually understanding the material." In these responses, we found a sense of surprise at past success with science rather than a sense of accomplishment.

The most impassioned responses to the question of a time the girls felt success with science came from the girls who had experienced hands on projects or experiential science learning. For example, two of the girls wrote about winning science fairs in elementary school. Anita wrote, "A time I had felt successful in science was at my fifth-grade science fair. I did an experiment about Coke and Mentos. Surprisingly, I won first place! I really felt great that day." Josephina wrote, "When I did my science project for science fair, I made a glow in the dark dry ice bubble and it made me happy because I ended up winning and then I started to like science! I started getting my materials and I wrote a conclusion and hypothesis and facts as a first step and I made a model and a poster board to explain everything." Coco wrote about a Spanish presentation she created outside of school with a classmate that was science related. "A time I felt successful with science was out of class for a Spanish presentation. My friend Nick and I had created dried ice bubbles. We combined water, dish soap, and dried ice into a 100ml beaker to create a reaction. We combined science with another language and we were successful!" Another student shared the importance of visiting the university during a summer scholars program for high school student and she had a chance to dive into research. "I had to research about many terminal diseases. It was something I loved. I had a project and I aced it. This was a great success to me because I discovered a career I would love to do." When we read these written responses, we saw all of these factors as teaching opportunities. Perhaps, if the girls in our project learned to see the causes for disconnect as external and malleable factors instead of internal and innate, then we could support more connection to science, more sense of control over how they saw themselves in relation to science, and more chance they would see challenge, or even failure, as part of a process of engaging with, learning from, and succeeding with science.

Building on Success—Daring to Dream

Through our GSW literacy community, we created a space where it was safe for girls to talk, write, and think about science. At the end of the first class, we asked the girls to reflect on what they had learned that day by writing reflections quickly on sticky notes. Even on the first day, the girls felt safe and supported enough in this room to share their connections, feelings, and thoughts in relation to science. In these brief, informal reflections, the girls also demonstrated how, even in the first day, the act of dreaming into science they began to shift their perceptions of themselves in connection to science. They started to move beyond times when they were left out of science and grow from their successes.

In the sticky notes, students shared the ways the literacy community allowed them to reconnect to prior experience, to themselves, and to science. For example, Anita wrote, "This program opened my eyes to see that we could be successful if we really try. This program is a nice way to really connect with ourselves." The girls also expressed thinking differently about science assumptions. Some of these had to do with ways that scientists look and things they do. For example, Josephina commented, "Another thought was how different my science teacher dresses compared to how I thought." She realized through the drawing a scientist activity that her teacher was a scientist and teacher. Josephina also realized that science was not disconnected from other fields when she said, "It takes a lot of thought in it. You have to write in science." In beginning to think differently about science assumptions, Josephina and the other girls opened up opportunities to dream or to think differently.

The girls also shared how the GSW community felt safe because it was made up of all girls. Ana wrote, "I should be more not afraid to speak, because we're all girls in science." And, Leah reflected, "I felt comfortable with all girls." Alicia wrote, "I learned how much so many girls like science." The girls were not accustomed to being in a classroom with all girls and it felt new and exciting. They noticed that being in a room full of girls allowed them to think about themselves as girls in connection to other girls. Students felt supported and noted how this support made them feel new possibility. For example, Estrella wrote, "I haven't been trying my best lately. And I know that I will have my science teacher and capstone teacher there to help me out whenever I need her."

The first day of the GSW gave the girls the opportunity to dream themselves into science. Bianca shared how the process of remembering her past experiences in relation to science allowed her to "learn more about myself" and how it also inspired her to return to a prior intention she had set for herself, "I remembered to stay determined with everything you do." Alejandra wrote, "I thought about how fun and exciting it would be to be a scientist. I also thought about the time in my advanced program we talked about women scientists." The GSW invited the girls to come together as a community around a shared topic and goal. It invited them into a supportive and focused community where they were allowed to dream.

Invitation to Dream: Integrating Writing and Technology in Your Classroom

One of our goals in the design and implementation of the GSW project was to bring women and science to this group of girls in as many ways as possible. We wanted to expand the girls' understandings of what it means to be a woman in science and to show them diverse examples. We integrated technology into each of our workshops in the form of TED talks, recorded interviews, and audio recordings of interviews.

To find resources, we did our own research. We pulled from things we had remembered hearing or reading, and we carefully and strategically looked for resources that connected to the girls' interests. We pushed the boundaries of what counts as science. We worked to find resources that upended gender stereotypes. We intentionally featured only women scientists. We also tried to find examples that were fun and silly. The invitation below is one of those examples.

In, "Why You Should Care About Whale Poo" Asha de Vos shares her research as a marine biologist, she specializes in working with marine mammals and creates mathematical models to protect Sri Lankan blue whales from ships in the ocean. In the TED talk technology invitation, de Vos shares a TED talk about science and the broadness of it.[1]

TED Talk Reflection

Instructions: After watching the TED talk about whale poo, think about the woman scientist who was speaking and answer the following:

1. What questions do you have for her?
2. What did you learn about science that you didn't know before?
3. What do you want to know about her path as a scientist?
4. What do you want to know about her personally?

1 www.ted.com/talks/asha_de_vos_why_you_should_care_about_whale_poo?language=en

Featured Women in Science: Resources for Teaching and Learning

Jane Goodall: Primatologist, Ethologist, and Anthropologist

Jane Goodall, a British primatologist, is famous for her 45-year study of wild chimpanzees in Tanzania and extensive conservation work through the Jane Goodall Institute. Jane Goodall Institute: www.janegoodall.org/

Autobiography

Goodall, J. (2010). *Through a window: Thirty years with the Chimpanzees of Gombe*. Boston, MA: Houghton Mifflin Harcourt. (Recommended for grades 8–12.)

Biography

Meltzer, B., & Eliopoulos, C. (2016). *I am Jane Goodall*. New York, NY: Penguin Books. (Recommended for grades 2–5.)

Children's Books

Edwards, R., & O'Brien, J. (2012). Who is Jane Goodall? New York, NY: Grosset & Dunlap.

McDonnell, P. (2011). *Me . . . Jane*. New York, NY: Little, Brown Books.

Ted Talk: What Separates us From Chimpanzees?

Retrieved from www.ted.com/talks/jane_goodall_on_what_separates_us_from_the_apes

Interview

Stromberg, J. (2013). Interview: Jane Goodall on the future of plants and chimps. *The Smithsonian Magazine*. Retrieved from www.smithsonianmag.com/innovation/interview-jane-goodall-on-the-future-of-plants-and-chimps-24872217/

Asha de Vos: Marine Biologist, Educator, and Conservationist

Asha de Vos's research focuses on Sri Lankan Blue Whales.

Ted Talk: A New Approach to Saving the Whales

Retrieved from http://voices.nationalgeographic.com/2014/10/07/a-new-approach-to-saving-the-whales/

Blog

de Vos, A. (2014). A new approach to saving whales. *National Geographic Ocean Views*. Retrieved from http://blog.ted.com/fellows-friday-a-magical-oceanic-feeding-frenzy-in-monterey-bay/

Interview

(2015). *National Geographic weekend interview*. Retrieved from https://video.nationalgeographic.com/video/til/161216-sciex-til-asha-de-vos-whale-poop

3 Learning to Ask

> One of my biggest challenges in becoming a scientist is learning you don't know everything and admitting what you don't know by asking dumb questions. I still struggle with this every day, but I also know everyone does. Questioning is not comfortable for most people and it is something you can practice and get better at.
>
> —Sarah, environmental quality manager

Four ninth and tenth-grade girls circle around Christina. One of the girls is holding a copy of *The Immortal Life of Henrietta Lacks* written by Rebecca Skloot (2011). The girls ask Christina about the book. Who is Henrietta Lacks? Who is Rebecca Skloot? Why did she write this book? Christina begins to explain,

> Henrietta Lacks was an African American woman living back east who had an aggressive form of cervical cancer and she ended up dying from it. However, doctors found the way her cancer multiplied was aggressive and remarkable. The doctors took samples of her cells upon her death. They kept doing research on the cells because of the way they multiplied. Her cells have been used all over the world for research.

Christina explains Henrietta's story, and then says, "She had cervical cancer. Do you know what that is?" The girls shake their heads no. "Do you know what a cervix is?" Again, no. For the next few minutes, Christina explains what a cervix is to the girls. This is part book talk, health education, and information about science writing. The girls listen intently. When they are done, they begin writing on their papers or move around the room to other posters and books. As they move, they read, write, talk quietly, and giggle some.

In this example, Christina and the girls asked questions and worked together to find answers. This process of asking was fundamental in GWS. In schools, the question-asking process is fundamental, especially in the English Language Arts. Teachers use Socratic seminars (Copeland, 2005), sentence starters (Olson & Land, 2007), Question Answer Relationships (QAR) (Raphael, 1986), and a variety of other strategies to encourage students to ask questions, find answers, and make meaning.

While students are often trained to ask questions in school, literature about the workplace shows women may ask questions in relation to work processes, but often neglect to ask questions that propel them forward in their careers, such as asking for a raise or asking to be considered for a promotion. Books such as *Women Don't Ask: The High Cost of Avoiding Negotiation and Positive Strategies for Change* (Babcock & Laschever, 2007) and *Nice Girls Don't Get the Corner Office: Unconscious Mistakes that Women Make That Sabotage Their Careers* (Frankel, 2014), tell the stories of women who fail to secure raises, promotions, prime offices, because they experience gender discrimination or do not practice specific skills to ask for what they want or have earned in the workplace. While these books are often oversimplified and tend to place the blame on women for the structural inequalities of the workplace, they do emphasize an important point: Learning to ask for what we want to know, what we need, or what we have earned is instrumental in gaining access and success in school, the workplace, and the community.

Feminism and Asking

As previously mentioned, one of the core characteristics of a community of practice is a shared domain (Wenger, 2006). In this case, the domain we created was a shared interest in women advancing in science through literacy. This domain is largely influenced by the fact women are underrepresented in STEM and women who do succeed often face barriers. As such, one of the core theoretical approaches we employed in this community of practice is feminist. We drew from Ryan, Myers, and Jones's (2016) concept of ecological feminism which, "recognizes the diversity of women's experiences within the shared experience of patriarchal oppression and acknowledges that 'values, notions of reality, and social practices are related'" (p. 9). In GWS we built upon these shared experiences of oppression to encourage the girls to ask questions and to ask for what they need.

We acknowledge a traditional definition of feminism only considers the oppression of women within a patriarchal society and does not fully capture the barriers to science careers many of the Latina young women at Metro Center Academy face. At Metro Center, the experience with feminism was far more intersectional. Intersectional feminism, as coined by Crenshaw (1989,

1991), accounts for the intersection of race and gender. At Metro Center, we considered the intersection of gender, race, and class in our feminist space. While women have achieved increased access to science fields, white women have largely achieved these gains. However, as Espinosa (2011) found, women of color who persist in STEM do so because of supportive communities. Therefore, in GWS we built on the principles of intersectional feminism to acknowledge and explore the factors limiting girls' access to STEM (gender, race, class) and in response build a supportive community of practice to increase access.

Even within this supportive, feminist community of practice, the process of learning to ask was still challenging and, at times, uncomfortable. Ryan et al. (2016) discuss this discomfort from a rhetorical perspective in relation to the rhetorical concept of ethos. They note that within a patriarchal (and racist and classist) system, "it is culturally and socially restrictive for women to develop authoritative ethé, yet . . . space can be made for new ways of thinking and artful maneuvering" (p. 2). As Ryan, Myers, and Jones passage suggests, it would be impossible for the participants in GWS to simply know how to ask for what they need, since social roles and practices have encouraged making nice and accommodating others. Therefore, we strategically scaffold the activities described in this chapter to make space for what Ryan, Myers, and Jones refer to as "new ways of thinking and artful maneuvering" (p. 2).

In this chapter, we describe the ways we structured the literacy curriculum in GWS to encourage asking questions of a new subject area, of new ideas, and of ourselves in connection to science. To do this, we designed the GSW workshop to push back against traditional notions of classroom-based writing and reading as formulaic and structured (Applebee & Langer, 2011; Wiley, 2000) to invite students to consider ways literacy is epistemic, or knowledge building. We used writing as a means for students to make connections with one another, with scientific voices via a variety of texts, and with scientific women outside of the classroom. These connections would aid students in learning what and how to ask within and beyond this community of practice.

A Scavenger Hunt

The anecdote that opens this chapter is from the Women in Science Scavenger Hunt activity, which invited students to use reading and writing to learn about women in science. In this activity, we created posters with pictures of women scientists and their biographies and placed them around the classroom for the girls to view and engage with (see Figure 3.1). The posters included women scientists from a variety of backgrounds with a diverse collection of expertise and accomplishments. For example, one poster described the work

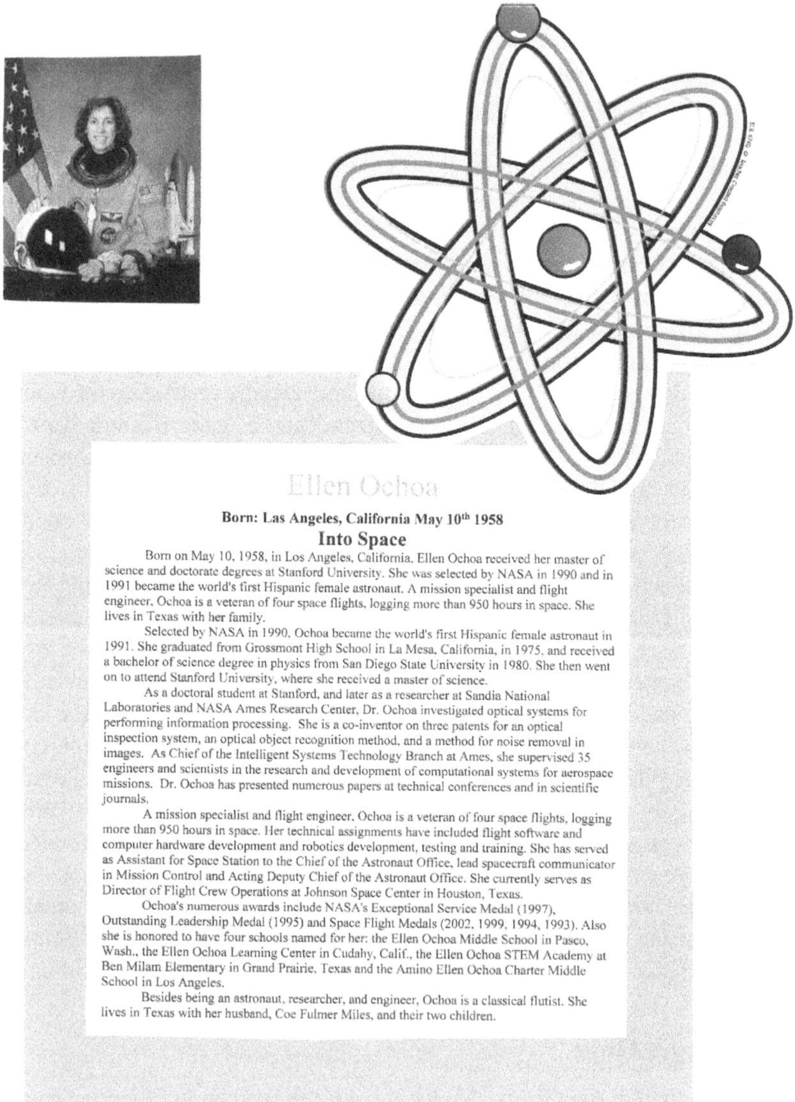

Figure 3.1 Women in Science Book Poster

of Rosalind Elsie Franklin. Franklin was responsible for much of the research and discovery work that led to the understanding of the structure of deoxyribonucleic acid, DNA. Another poster described Helen Sawyer Hogg, an astronomer well known for her research on variable stars in globular clusters.

She was also a writer and remembered for her astronomy column, which ran in the Toronto Star from 1951 to 1981. We found information about women in science to use in creating the posters by doing web searches for women in science. We also drew from a number of women and science resources, such as Ignotofsky's (2016), *Women in Science: 50 Fearless Pioneers Who Changed the World.*

Along with the posters, we placed books about women in science around the classroom for the girls to peruse (see Table 3.1). The books were a mix of children's books, autobiographies, and biographies. Students were encouraged to engage with these texts in a variety of ways: walking around to numerous stations, reading extended excerpts of texts, talking with other students or us about the texts, etc. As they made their way around the classroom, we invited the girls to look for and write about: 1) a woman whose work you admire, 2) a woman whose work is interesting, but you don't fully understand, and, 3) a woman whose work you know about but want to know more.

One of the explicit goals of the Scavenger Hunt activity was for students to use reading and writing to think more broadly about women in scientific fields. Another goal was for the young women to practice the skills connected to asking by using questioning, talk, and reflection in connection to learning about women in science. Some girls sat on the floor reading, while others made their way around in small groups. In these small groups, girls talked about what they saw, asked each other questions, and wrote reflective and informational notes in response to the materials.

The Scavenger Hunt activity allowed students an accessible, interactive, and engaging way to learn about material through a variety of texts and written responses to texts. This can be enacted with analog or digital texts (for more ideas on how to use scavenger hunts see: Eagleton, Guinee, & Langlais, 2003; Fleischer & Andrew-Vaughan, 2009). As the girls read, wrote, and reflected on the lives and careers of notable women in science-related fields, we asked them to begin thinking about what they knew or did not know, about issues of access and equity in relation to science fields, and about the ways women develop grit and determination. Through the questioning and interrogating of texts in the book hunt, the girls began to position themselves in conversation with notable women in science. They began to connect to their lived experiences, stories, successes, and failures. Additionally, the girls used this activity to relate or express interest in the work of the women scientists. For example, Amelia, one of the girls who participated in the book talk about *The Immortal Life of Henrietta Lacks*, wrote about her interest in the book and in Lacks's story, "I chose this book because this woman had a type of cancer and I am interested in cancer." For Amelia, who was interested in oncology because a family member had cancer, Skloot's writing and Lacks's story were particularly interesting because "Her cell are important for medicine." Amelia was one of the girls who took part in the opening anecdote.

Table 3.1 Women in Science Book List

Title	*Author*	*Age Group*
Amelia to Zora: Twenty-six Women Who Changed the World	Cynthia Chin Lee	Children's
Annie Jump Cannon, Astronomer	Carole Gerber and Christina Wald	Children's
I Wonder	Annaka Harris and John A Rowe	Children's
Look Up! The Story of the First Woman Astronomer	Robert Burleigh	Children's
Marvelous Mattie: How Margaret E. Knight Became an Inventor	Amelia Arnold McCully	Children's
Mary Anning and the Sea Dragon	Jeannine Atkins	Children's
Me . . . Jane	Patrick McDonnell	Children's
Rachel Carson and Her Book That Changed the World	Laurie Lawlor and Laura Beingessner	Children's
Rosie Revere Engineer	Andrea Beaty	Children's
Sofie Scott Goes South	Alison Lester	Children's
Summer Birds: The Butterflies of Maria Merian	Margarita Engle and Julie Paschkis	Children's
The Fossil Girl: Mary Anning's Dinosaur Discovery	Catherine Brighton	Children's
The Most Magnificent Thing	Ashley Spires	Children's
The Watcher	Jeanette Winter	Children's
Violet the Pilot	Steve Breen	Children's
Who Says Women Can't Be Doctors?	Tanya Lee Stone	Children's
Girls Who Looked Under Rocks	Jeannine Atkins	YA
Temple Grandin: How the Girl Who Loved Cows Embraced Autism and Changed the World	Sy Montgomery	YA
The Evolution of Calpurnia Tate	Jacqueline Kelly	YA
The First Woman Doctor: The Story of Elizabeth Blackwell, MD	Rachel Baker	YA
Ada's Algorithm: How Lord Byron's Daughter Ada Lovelace Launched the Digital Age	James Essinger	Adult
Barbara McClintock: Pioneering Geneticist (Makers of Modern Science)	Ray Spangenburg and Diane Kit Moser	Adult

Chrysalis: Maria Sibylla Merian and the Secrets of Metamorphosis	Kim Todd	Adult
Grace Hopper and the Invention of the Information Age	Kurt Beyer	Adult
Hypatia of Alexandra: Mathematician and Martyr (Astronomer as Well)	Michael A. B. Deakin	Adult
Lise Meitner: A Life in Physics	Ruth Lewin Sime	Adult
Nobel Prize Women in Science	Sharon Bertsch McGrayne	Adult
Rita Levi-Montalcini: Nobel Prize Winner	Susan Tyler Hitchcock	Adult
Rocket Girl: The Story of Mary Sherman Morgan	George D. Morgan and Ashley Stroupe	Adult
Rosalind Franklin: The Dark Lady of DNA	Brenda Maddox	Adult
Rosalyn Yalow: Nobel Laureate Her Life and Work in Medicine	Eugene Straus	Adult
The Fossil Hunter	Shelley Emling	Adult
The Sound of a Wild Snail Eating	Elisabeth Tovah Bailey	Adult
The Comet Sweeper: Caroline Herschel's Astronomical Ambition	Claire Brock	Adult
The Girls of Atomic City	Denise Kiernan	Adult
Through a Window: Thirty Years with the Chimpanzees of Gombe	Jane Goodall	Adult

She did not know what cervical cancer was. However, through her questions during the book talk and her engagement with the text about Skloot's book, Amelia had expanded understanding about a medical field of interest. In this case, Amelia asked questions and received answers that tied together a family experience, gave her newfound knowledge about cells, and engaged her in a new experience with a text about science.

Mariana, who read about Lise Meitner during the Scavenger Hunt activity, saw the problem in Meitner's brother taking credit for her work. She also noted similarities between herself and the scientist, writing, "Meitner knew physics and her brother knew chemistry. They worked together on radioactivity. I admire this woman because I enjoy physics and chemistry." Mariana used this writing and reflection activity as an opportunity to make connections between the interest of a notable woman scientist and her own interests. Mariana's initial questioning process helped her identify the sexism prevalent in scientific fields and focus on her deep respect for women, like Meitner, who persist despite facing gender barriers.

In their explorations during the Scavenger Hunt, many of the girls questioned how the women scientists broke through barriers or limitations. For example, Alicia a student who read about Ellen Ochoa, the first Latina woman astronaut, asked, "What were the obstacles you had to overcome to get where you are? What were the struggles women overcame to get into science fields, but also her personal struggles?" Alicia's questions represented a trend in this reflective writing—an understanding that barriers to women's science careers exist. Mariana, another student, wrote that Lise Meitner "knew physics and her brother knew chemistry. They worked together on radioactivity. I admire this woman because I enjoy physics and chemistry." However, Mariana went on to question, "Was she bothered by her brother getting all of the credit?" In the majority of responses, the girls wrote about barriers, boundaries, others who took credit for the work, etc. and showed a profound understanding of the barriers for women in science-related fields. The opportunity to use writing to express feelings and ask questions about these barriers is significant in a feminist approach to reading, writing and research. For the girls to see themselves in conversation with other women in science, they had to understand issues such as the systemic exclusion of women in these fields.

In this activity, the girls practiced asking questions and finding answers to explore the lives of notable women in science. Many of the girls came to see themselves in conversation with notable women in science. They also came to see many of the struggles women scientists experience as they gain legitimacy in their fields of study and made connection to the work of these women. In doing so, they could see themselves in this work and envision this as part of their future lives. One student, Laura, reflected on the activity saying, "It's good to learn about a variety of topics even if you won't go into it in the future. Maybe I will." Laura's understands that gaining a breadth of knowledge and reflecting on it opens up possibilities for her in ways she may not have been aware existed. Laura's comment "Maybe I will" reflects this potentiality.

Asking to Understand

Building on the questioning in the Scavenger Hunt activity, we wanted the girls to start thinking about the women in science they wanted to interview. To begin this process, Jessica invited students to brainstorm areas of interest in science and respond to the following prompt: "Write a brief note sharing what area(s) of science interests you and imagine what kind of scientist you might like to be one day. What kind of scientist would you like to interview and why? How is this profession connected to a life path you can see yourself pursuing?" Using writing to articulate, imagine, and brainstorm interests in science allowed students to articulate their interests, whether clear or fuzzy. This note format was important because it offered a different way for the girls to experience schooled writing—as informal and needs based. Via the note, the girls were invited to express interest, need, and a vision for their futures. In essence, the note format encouraged the girls to ask for what they need in as specific a way as possible within a supportive feminist space.

Some students came to the project because they were interested in science and already envisioned their future lives in connection to science. For example, Amelia's process of asking was clear and direct. She had future goals, and, through this writing process, she asked to be connected to a woman pediatric oncologist who could help offer her knowledge and advice in pursuit of those goals. In her reflection, Amelia wrote, "I want to interview an oncologist because I want to become a pediatric oncologist in the future. I have always found cancer so interesting. I have also had cancer in my family." Another student, Clara, described her interest in interviewing a botanist. She had specific things she wanted to gain from her interview in terms of access to information and a better understanding of this career path. "I want to be a botanist in the future, so this interview can help me with my future job. Also, the woman I interview can help me understand what is required for a degree in botany." For 14-year-old Estrella, a freshman, imagining herself in a future role as a scientist was not a problem. She wants to become a neurologist. "My uncle got into a car crash and had some problems after that, so I want to help him," Estrella said. "I want to figure out why a person reacts a certain way to this environment. . . . I want to see the reactions." For sophomore Coco, 16, the idea of becoming a cardiac surgeon sounds enticing. She has her heart set on attending medical school in a few years. "The first time I got interested in science was back in middle school when we dissected a baby shark," Vargas said. "From that point on, I really wanted to cut things open and see what's inside."

Some students asked for help homing in on a science-related interest/career path to narrow overly broad visions, such as Mariana's, "I'm very fascinated in chemistry, physics, engineering, and sustainability. I'm not sure which to choose?" Other students had never thought about how their interests were connected to science before and came to the project because teachers had suggested it would be worth their while or because they had a friend who was interested, and they wanted to stick together. For these students, imagining their future lives or interests in relation to science was challenging because it was not necessarily something they had considered before.

Another trend we saw in the notes was a sort of scientific wondering. That is, a number of the girls asked to be connected to women in more than one field and these fields were often divergent. For example, Alejandra wrote, "What I think I'm going to do is interview a person who is in marine biology or an architect because I love marine animals and I love to draw. I wanted to work at Sea World because of the animals and an architect because of the many ideas I have about buildings and I really enjoy drawing." This idea of wondering is important in learning to ask. Alejandra understood that the space was safe and supportive, and she could ask for what she wanted and needed, but she also demonstrated her ability to use writing to think and wonder as a way to clarify her thoughts. The First Thoughts Note was simply that—her first thoughts. In GWS, we wanted the girls to feel free to ask and explore without the pressure of polishing or perfecting their thoughts while pursuing areas of interest and strength, both of which Alejandra expresses here. Girls with specific goals and visions of their future—girls who knew what to ask for—and girls with less specific goals and visions—those who did not know exactly what to ask for or whose ideas diverge—brought equal value to this process. We shared with the girls how asking for what you want is as important as using a process of asking questions to figure out what you want.

The girls also used the First Thoughts Note to express their questions about ways of accessing science careers and the conditions for women in these careers. Clara wrote that she wants to be a botanist and would like to interview a botanist to ask, "What things are needed or required for a degree in botany." Additionally, Clara wanted to hear "stories about how she traveled somewhere to understand the plants' habitats." In her process of asking, Clara expressed an interest in understanding how to access her field of interest, but also in knowing how knowledge is made in that field using research. Anita asked to interview a doctor. More specifically, Anita wanted, "To know more about her path in getting into what she has become today. To learn all the challenges she faced." Anita hoped to understand about the path and challenges, "to see if I truly want to be like her." In this letter, Anita expressed an interest in asking about the paths and challenges of women in a field in

Figure 3.2 Christina works with a group of students as they write

Source: Charlie Leight

which she hopes to work. Anita's asking shows awareness of the challenges for women in her intended field and an interest in understanding whether she wants to take on those challenges herself.

The First Thoughts Note gave the girls the opportunity to use writing to ask for what they wanted and needed. After these notes were complete, we talked with each of the girls about their choices. For example, we asked Alejandra whether she would rather talk to a marine biologist or an architect to begin placing the girls with interview subjects. This gave us an opportunity to ask questions of the girls and to further establish the importance of using literacy to ask for inquiry, knowledge, and access. This also served a very practical skill for us by giving us information to begin building a network of women scientists for the interviews.

Strategies for Asking

Once the girls had been invited to ask for what they needed via the First Thoughts Note, we invited them to build upon these skills in the process of writing practice interview questions. At this point, the girls were ready and excited to use their literacy skills to gain access to the knowledge and experience of professional women working in fields they wanted to explore. This chapter focuses on the ways the girls learned to ask and gain access to information about science careers by writing interview questions. In Chapter 4 we discuss the process of building a network of women scientists for the interview process.

At this point, the girls participated in the first part of a three-part process. Part one will be described here, since it focuses on experience with asking and parts two and three will be discussed in Chapter 4. For part one, Jessica taught a mini-lesson on writing interview protocols to give students tips for creating successful questions to garner rich, varied, and helpful responses in their interviews. She began by asking students to imagine sitting across from their interviewee at a coffee shop, "What questions would you ask this person over coffee?" Next, we showed the group a Ted talk by a marine biologist, Asha de Vos (2014), "Why You Should Care About Whale Poo," and, after listening to this engaging and dynamic scientist, students wrote in response to the following questions:

a. What questions do you have for her?
b. What do you want to know?
c. What do you want to know about her path as scientist?
d. What do you want to know about her personally?

Jessica gave students these examples of open-ended questions to use as a starting point for creating their own protocols. Amelia wanted to know the impact of this work on the person, "How has this work changed you?" Other students wanted to know about the scientist's motivation for doing this kind of work, "What motivated you to learn about whales and how they affect the ocean and us as a species?" Other students wanted to know about her pathway to becoming a scientist, "How long did it take her to get where she is? When did she decide she wanted to study whales?"

We encouraged students to frame their interview questions to help them garner useful information in imagining, reflecting, or moving toward their future life paths. This lesson, and corresponding writing activities, helped the girls to learn rhetorical skills and strategies and specific genre conventions typically used in interview questions and processes to gain access to knowledge about women in science careers. Furthermore, in discussing interview questions we covered oral and written interviews, since many of the girls planned to conduct email interviews. This first step further developed the girls' ability to and comfort with asking in preparation for their interviews.

In this chapter, we describe a scaffold approach to using literacy activities to learn to ask within a community of practice and built on feminist principles. We acknowledge that learning to ask is challenging, especially for young women who may have learned to accommodate others, rather than assert their needs. However, by engaging with texts, writing informal notes, writing interview questions, and taking part in peer response, the girls in

GWS had many different experiences with asking questions and community building. Through these questions they experienced literacy as a socially constructed and supported act (Bakhtin, 2010) communicated with scientists and historical figures, the researchers and teachers, and, most importantly, each other. Within this community of practice, the girls built confidence in their literacy skills and learned to use those skills to ask the questions they needed to gain access to future lives.

Invitation to Ask: Using Peer Response for Questioning and Community Building

One of our goals in GWS was to invite students to further develop their literacy skills within a supportive community of practice. However, this commitment to using literacy skills for community building is not exclusive to GWS.

It is part of *all* of our writing classes. One of the primary ways we build community via literacy skills is through self and peer response to writing. Take note: we are calling it response, not editing or review. In self or peer response we encourage students to engage with their own or with one another's texts, write about those texts, and talk about those texts. This is more than surface level editing or checking boxes on a checklist, this is a way of engaging deeply with texts to make significant, audience-focused changes to a text. When students write knowing that readers will engage with their text in a meaningful way, they come to see themselves as authors, rather than just students turning in papers for a grade.

Here is an example of a self-guided review for students:

Step 1: Spend 5 minutes reviewing all of the notes, outline, and brainstorming you have written so far. Let your head swim with information.

Step 2: Now fast write about your essay topic for 8 minutes. Tell the story of how your thinking about this topic has evolved.

Step 3: Skip a few lines on your page. Now fast write for another 5 minutes, this time focusing on more specific case studies, situations, people, experiences, observations, and so on that stand out in your mind about this essay topic. Keep your pen or pencil moving for the whole 5 minutes.

Step 4: Finally, skip a few more lines and write these two words in your notebook: "So What?" What is your news? What is your main idea? Why does it matter?

Invitation to Ask: Resources for Incorporating Peer Response in Your Classroom

Here are some resources to help incorporate effective peer response in your classes.

Articles

Saidy, C., & Early, J. S. (2016). You need more organization bro: Relationship building in secondary writing and revision. *The Clearing House: A Journal of Educational Strategies, Issues, and Ideas for Middle & High School Teachers*, *89*(*2*), 54–60.

Early, J. S., & Saidy, C. (2014a). A study of a multiple component feedback approach to revision for secondary ELL and multilingual writers. *Reading and Writing*, *27*(6), 995–1014.

Early, J. S., & Saidy, C. S. (2014b). Uncovering substance: Teaching revision in high school classrooms. *Journal of Adolescent and Adult Literacy*, *58*(3), 209–218.

Books

Harris, J. (2006). *Rewriting: How to do things with texts*. Boulder, CO: Utah State University Press.

Heard, G. (2014). *The revision toolbox: Teaching techniques that work*. Portsmouth, NH: Heinemann.

Online Resources

National Writing Project. Retrieved from www.nwp.org/cs/public/print/resource_topic/revision

4 Learning to Network

> I guess I've learned that science is even more important than you think. I knew that, but it's still. . . . You keep learning even more every day about science. It's pretty interesting.
>
> —Alexandra, tenth grade

Networks and Sponsorship

In Chapter 3, we explored the ways the girls learned to ask, or advocate for their interests and needs as they continued their exploration of scientific fields and careers. Based on those needs, one of the goals of this project was to create a strong network of students, women in science, teachers, and researchers. We hoped that this network would be both supportive and sustainable—supportive because we hoped to identify women in science interested in increasing access for Latina girls to scientific fields, and sustainable, because we wanted these networks to provide support for the girls beyond this initial project.

Espinosa's (2011) article "Pipelines and Pathways" explored the persistence and achievement of women of color in college STEM programs. For women of color, specifically, Espinosa found that both a peer system of support and clear mentoring added to the persistence in STEM fields for women typically underrepresented and often excluded from these fields. Our project, builds on two of Espinosa's specific findings. First, Espinosa found women in STEM who pursued peer relationships in their fields, especially via STEM focused clubs, were more likely to persist. We saw GWS as a high school version of one of these clubs encouraging peer interaction. Our meetings, which functioned somewhat like a club, brought girls with a common interest together to explore their interests, discuss and challenge existing boundaries to their participation, and especially offer support in their pursuits of STEM interests. Espinosa's second finding, on which we built, is that women of color pursuing STEM are often positively impacted by participating in structured research activities. The positive impact is, in part, due to the activities themselves, but is also due to the fact that research programs often "provide role models and avenues for continued science performance" (p. 232). In GWS we combined the peer element with role models who could share avenues for success in the sciences. In doing so, we built a supportive network of girls and women interested in STEM.

The second element of this network is women who provide examples of success, information and experiences of limitations in STEM, and encouragement via personal experience. In thinking about the role of these women, we draw from Deborah Brandt's (2001) description of literacy sponsors. Brandt notes that sponsors, generally, "are powerful figures who bankroll events or smooth the way for initiates . . . sponsors nevertheless enter a reciprocal relationship with those they underwrite" (p. 19). In regard to literacy, more specifically, Brandt acknowledges that sponsorship has many forms that include, "a range of human relationships and ideological pressures that turn up at the scenes of literacy learning—from benign sharing between adults and youths to euphemistic coercion in schools and workplaces . . ." (p. 20). Research has shown how successful writers often have a person or institution in their lives sponsoring their acquisition of writing skills and writing growth is related to access to quality writing mentors and rich learning contexts (Brandt, 2001; Early & DeCosta, 2012; Fishman, Lunsford, McGregor, & Otuteye, 2005; Yosso, 2005).

The concept of literacy sponsorship is apt in describing the relationship between the students and the women in science in GWS since the women who agreed to be interviewed and share their expertise were sharing their professional science literacy as well as their personal experiences with the girls in an effort to expand the experiences of the workplace. As we developed the curriculum, the concept of science literacy sponsors helped us to consider ways that writing and literacy activities are used to approach people with more expertise and professional stature. We kept this idea of sponsorship in mind as we built the network and structured literacy activities that helped the girls access their sponsors' knowledge.

Connections and New Networks—Reaching Out to Women in Science

The building of this network was both challenging and rewarding. As mentioned in Chapter 3, the girls began by composing a "First Thoughts Letter." In this First Thoughts Letter, they wrote about what type of scientist they would like to interview, what they wanted to know from this scientist, and what they hoped to gain from this learning. The list was broad, and we set out to pair girls as closely as possible with women scientists in the fields they wanted to explore.

In terms of building this network and providing access to women in science, students first had the option of drawing from their own social networks.

Two of the girls chose to interview a science teacher they love at Metro Center Academy. Interestingly, this teacher was preparing to leave teaching to attend medical school, so she provided a resource about science teaching and medical careers. Another student interviewed her cousin, a psychologist.

If students did not know women scientists to interview, we paired girls with women scientists doing work connected to the girls' interests. We set out to find women in science by emailing professors we knew through work, reaching out to mothers of our children's friends, and cold calling public and private community businesses. The women who agreed to be interviewed were thrilled to be of help and expressed enthusiasm and interest in the project. For example, when Christina reached out to a pediatrician by email, she responded, "I would love to participate!" A woman who worked for the state's department of environmental quality told us how excited she was to participate in the project, and how sincerely committed she is to increasing access to science fields for women.

This pairing was not without challenges, however. One major challenge was finding women working in fields in which we had no contacts. For example, a few of the girls wanted to interview a forensic scientist, as they had great interest in this type of work. However, neither of the teachers, nor Jessica and Christina, had contacts in this line of work. Christina began by contacting a male friend who is a police sergeant. He put her in touch with a woman in the city's forensic crime lab. However, forensic scientists are extremely busy and this one was difficult to contact. In the meantime, Christina continued exploring options for other forensic scientists in the area. A neighboring city was excited to provide the students with access to a forensic scientist and to her lab. However, since the forensics lab was so busy, they wanted the students to visit the lab, which was not possible for the girls who did not have transportation. These types of challenges were present in a number of the other pairings as well. For example, we had no trouble locating pediatricians and midwives who were open to being interviewed. We both have young children and we used our personal contacts, our children's pediatricians, and friends from work and the community. However, we faced more challenges finding a woman pediatric oncologist to be interviewed. In this case, Christina cold called local hospitals and spoke to their public relations offices to try to locate a pediatric oncologist who would be interviewed. Ultimately, one was not located, and the student interviewed another physician. This experience was important because the student learned that sometimes in networking she would need to be flexible and use the contacts she had.

In this stage of the project, in which students were paired with women scientists, Jessica and Christina did the majority of the work in establishing the network of girls and women in science. We used our contacts and our communication skills to help the girls gain access to women in fields of interest. In this case, we leveraged our positions of privilege as university professors, which teachers could do as well, to make these connections for students. While we understand that this could be viewed negatively, since the network did not develop organically, we hold that most networks do not simply develop. Rather, networks are cultivated in a variety of ways and individuals often receive support and guidance in accessing them. We also saw ourselves as part of the created network. We were deeply invested in finding and making these connections for the girls and we made this work of taking risks, reaching out, and building connections transparent to them. Christina made some of the calls to women in science from the classroom or left the room to take a call from one of the women scientists. The girls knew they could ask us for help and we would do our best to help them. Aja, who hoped to interview a psychologist, wrote, "If you guys know someone in that field, I would really appreciate if you can connect me with them." Like many of the other girls, Aja trusted and valued our adult ability to gain access that she could benefit from. As part of the teaching, we were honest about the ways we built the network, and when we were unable to pair the girls with women in the exact fields they wanted, we helped them to see how fields were connected and how women in a sub-field or connected field could still be a resource.

A final challenge we faced in this part of the project is that we had hoped that many of the women scientists included in the network would be Latina women, or women of color, so that the students would see themselves represented in the women interviewed. However, this was not the case. In fact, many of the women were white women. Of course, this confirms what we know from research about the representation of Latina women and women of color in scientific fields (Ong et al., 2011). However, this was one disappointment as we compiled the list of interviewees and built this network. We felt that if we had more time to build the network, we could have ensured more diversity in the interviewees.

Entering the Conversation—Writing Interview Questions

The process of matching the girls with women in the network was lengthy. Some of the girls were paired fairly quickly, while other pairings took up to two weeks. As is often the case in classrooms, we could not stop instruction because one element of the learning was not complete. As such, even while we were still working to develop the network, we began workshops to help the girls access the network of women sponsors via interview questions and emails. To further inform our choices, we polled the girls about their experiences with interviews and writing professional emails, and, on the whole, they reported little to no experience. This knowledge helped us to think about ways that we would sequence this instruction to most effectively help them gain access to the women interested in acting as sponsors.

Since we were still working to pair all of the girls with sponsors, we returned to the early interview skills the girls had learned in Chapter 3. Jessica then led a mini-lesson that invited students to refresh and expand their interview question knowledge. She started this workshop with a TED talk by Suzanne Simard (2016) called "How Trees Talk to Each Other." This specific TED talk was chosen because it was both interesting and a type of science that none of the girls had identified as an interest. In this TED talk Simard describes her interest in forestry and Canadian Forests. This helped to continue developing ours and the girls' perspective on "what counts" as science while also focusing in on the types of questions the girls might ask of a scientist who studies trees in Canadian forests, which was quite different than their experiences in the urban Southwest.

The students watched the TED talk, took notes, and discussed what they saw. Then, Jessica led the students in the workshop in which she provided a brainstorming activity that included three questions (see Figure 4.1). The sample questions were intended to offer the girls a way to begin asking interview questions and to help them think about ways to ask questions that would yield rich results. Building off of Jessica's starter questions, the girls then worked in groups to discuss the TED talk more and to think about what they would want to know about Simard and her work as a scientist. After their work in small groups, the girls formed questions of their own that they might ask Suzanne Simard. The girls expressed a deep interest and respect in Simard's work, which was reflected in their questions. For example, many of the girls wrote questions asking about Simard's experience of living in the trees. These questions ranged from Amelia's simple inquiry, "What inspired you to investigate trees and study them?" to Leah's more pointed question, "When you climbed those trees what did you like about climbing them, and what did it feel like?" The majority of the girls wrote questions focused on the ways that Simard developed her passion or interest. These questions included inquiries about when this interest developed. For example, Sara wrote, "At what age did you figure out you wanted to study trees," which focused on how scientific

Creating Interview Questions:

A Brainstorm

Let's create possible questions for our interviews together as a group and individually. Here are a few starting questions/ideas:

- Was there a turning point when you decided to pursue this career path?
- What advice would you give a young adult interested in this line of work?
- What is the most rewarding experience you have had in your work and what has been a challenge?

Figure 4.1 Creating Interview Questions

passions develop. Many of the girls wanted to know specific questions about Simard's career, so they wrote specific questions about Simard's schooling or experience. Sara's questions "Where did you study about trees? What type of school did you go to?" reflected the questions of many of the girls who wanted to understand the type of education that a sponsor would need to do this type of work. While many of the girls understood that the scientists they would be interviewing had all gone to college, they wanted to more deeply understand what that meant. What types of programs in college? How long does this take? How hard is it? This activity with Simard's TED talk brought these types of questions to the forefront and pushed the girls to think deeply about what it means to get an education as a scientist and the full range of experiences that education can include and the pathways people take to enter fields and careers.

As was the case with the Scavenger Hunt activity, described in Chapter 3, many of the girls wanted to understand the obstacles to scientific study. Clara asked, "What are some obstacles you had to overcome to get where you are now?" while Carla asked "Did you ever second guess your job? If so, why?" These questions reflected a deep understanding of the complexity of this scientific work and the challenges of doing work that is uncommon. Josephina asked, "Do you ever feel scared going up in those trees?" This pointed question reflects the ways that the girls were processing and trying to make sense of a science career that seemed new to them. While questions about setbacks, obstacles, and fear were present, there were many more questions focused on persistence and success.

In focusing on persistence, we saw a slight change from the original Scavenger Hunt activity. The girls were focused on obstacles and limitations in scientific work but also on the potential for this work to impact others, the researcher, and the field. Many of the girls asked about moments of pride in the researcher's work. For example, Blanca wrote two related questions: "1) What moment in your career life are you most proud of? 2) What was one of the best experiences you've had in this work?" This focus on moments of pride and persistence was present in a number of questions. Aja wanted to know how these moments of pride and persistence impacted others. In her questions she wrote, "1) Have you discovered something that you made everyone change their minds about? What was it? 2) What's the thing you love the most about what you do (your job)?" In these questions, the girls began to see the potential for questioning a sponsor. Learning about the true nature of work in the sciences was as much about understanding the education and training required as it was about understanding the sponsor's passions, successes, and achievements.

The final focus of the questions was on the types of sponsorship the scientist herself received. For example, Estrella wrote, "Who did you look up to when you were younger? How did it impact you then and now?" Miranda wrote "Who encourages you to keep up your research?" These questions show a fundamental understanding of a woman's need for support and sponsors to achieve and maintain success in scientific fields. For many of the girls, GWS was largely about using literacy to learn to see these structures of support in scientific fields. The girls also began to develop a very sophisticated view of sponsorship through this workshop. While they understood that emotional support was important, they also came to understand the need for financial and institutional support. Coco wrote in her questions, "Who helps fund your travels and studies?" This astute question shows Coco's understanding that sponsorship is more than emotional and experience sharing, it includes financial support that allows women to persist in their work.

This early interview question workshop gave the girls a safe way to begin thinking about the things they wanted to know from their interviews. By first writing to Suzanne Simard, the girls expanded their thinking about what it means to be a woman in science and posed questions for a real woman in science that would be both interesting to know and would help them shape their own questions later on.

Accessing the Network

Once each of the girls had practiced interview questions and were placed with a sponsor woman scientist, Christina taught two workshops specifically focused on entering the network of women in science and addressing the sponsors. The first workshop focused on composing interview questions for the specific sponsor and the second focused on writing professional emails. At this point, all of the girls had written some interview questions, and all had written emails before, at least to teachers at school if not to other audiences. Thus, these two workshops focused on honing those skills for the sponsor audience. Addressing a sponsor, whether in person or via a written genre such as email, can be intimidating for those being sponsored. Sponsors are often more experienced, busy, and older than those being sponsored. They are part of unfamiliar worlds. As Brandt (2001) reminds us, while sponsors often give of their time and expertise, this is not without some type of benefit to them. In the case of the women sponsoring these girls, many were excited to provide information and mentoring support that would provide greater access to their fields for women of color. However, the women sponsors were busy—most of them managed demanding careers and families—and appreciated professional communication that respected their expertise and time. These two workshops focused on helping the girls to access this professional network, which was new to them as high school students.

Advanced Interview Workshop

The advanced interview question workshop acknowledged that the girls already had experience with interview questions, so it moved beyond that to ask questions inviting the sponsors to share their expertise. To begin this workshop, the girls all started by brainstorming things they wanted to know from their sponsors. Christina then helped the girls to think about ways to invite the sponsors to tell their stories of success in science while answering the girls' questions. To do so, Christina provide six specific guidelines for open-ended interview questions that would help the girls expand their early experience with writing interview questions and encourage lively interviews:

1. Avoid yes or no questions.
2. Keep questions on topic.
3. Ask questions that cannot be answered by doing research.
4. Invite the sponsor to share a story or example.
5. Write questions to encourage the interviewee to expand on ideas.
6. At the end invite the interviewee to share something you did not ask that she might think is important for you to know.

These guidelines were provided to help the girls think about ways to shape their interview questions to yield the most information in the shortest amount of time. So, for example, the third guideline focuses on asking questions that could not be answered by doing research. We encouraged each of the girls to do a web search of their sponsor woman and look at her LinkedIn profile and any other public profiles. In doing so, the girls could find out where the sponsor worked, where she had gone to school, etc. They could read resumes and professional descriptions of the science work these women participated in. This also gave the girls a way of understand the sponsors and the public and professional ways the women scientists use writing to share their public identities and professional stature. After reading these online spaces, the girls created interview questions building on existing knowledge. This would allow them to use their limited interview time to ask more in-depth questions. Many of the girls used this information to their benefit. However, some of the sponsor women did not have professional profiles that could be accessed. For example, the forensic scientist, who worked for the largest police department in the country, did not have a public profile. We used this as an opportunity to discuss why a person who runs forensic tests and often testifies at trial may not have a public profile. This first research step helped with the interview questions, but also helped the girls to begin seeing both the risks and rewards of the careers.

The girls worked on their own interview groups in small groups and then paired for a quick peer review of the interview questions. In the peer review, each of the girls read her peers questions and then completed the Interview Question Checklist for her partner. The Checklist (see Figure 4.2) invited the girls to evaluate the interview questions based on the six criteria and to suggest revisions to the questions. The process of peer review solidified that the girls were all part of the network of women interested in science. They brought their knowledge to the network to help their peers achieve their goals and further their knowledge. As the girls completed the peer review, Jessica and Christina made their way around the room asking and answering questions and encouraging revision of the interview questions. We worked *with*

Peer Response Form

Interview Question Checklist

Directions: Listen to your partner's interview questions and answer all odd questions. Return the paper back to your partner so she can complete all of the even numbered questions (bolded) based on your feedback.

Your Name__________________ Checklist For _______________

1. Does my partner use yes/no questions? _______Yes _______ No
2. **Revise any yes/no questions.**

3. Are all of the questions on topic? _______Yes _______ No
4. **Rewrite off topic questions to make them more on topic.**

5. Does my partner ask questions she could answer by doing research? ______Yes ______ No
6. **Rewrite any questions that could be answered by doing research.**

7. Does my partner ask the interviewee to share a story or example? _______Yes _______ No
8. **Rewrite one question to invite the interviewee to share a story or example.**

9. Does my partner ask questions that encourage talking? _______Yes _______ No
10. **Revise one question to encourage the interviewee to talk.**

11. Does my partner offer the interviewee an opportunity to share something more? _______Yes _______ No
12. **Rewrite the last question to offer the interviewee the opportunity to share something you haven't asked.**

Figure 4.2 Peer Response Form

the girls in forming their questions and offered the type of feedback their peers were also offering.

Each of the girls ended the workshop with approximately 8–10 questions they wanted to ask their sponsors in science. As we analyzed the data, we saw some trends in the types of questions asked. Some of the common themes in the interview questions addressed obstacles and persistence, which were recurrent themes in GWS that started in the early Scavenger Hunt activity and carried through the project. These themes reflected the girls understanding that obstacles exist for women in science and that those who persist must overcome these obstacles. Yet, this was addressed in a variety of ways. For example, Leah asked, "What is an obstacle you overcame before you started this line of work?" Aja asked, "What was a major obstacle when choosing your career of psychology?" Mariana asked the question in a more pointed way, "What are some obstacles you faced when achieving your career goals? (ex. Gender? Stereotypes? Race? Etc.?)." These questions were focused on early obstacles the interviewees may have had to overcome, and as Mariana's question shows, a deep understanding that specific obstacles are particularly limiting for women in science.

Other girls asked questions more focused on the sponsor's persistence or satisfaction in her line of work. Carla asked her interviewee, "Did you ever second guess your job and why?" Miranda asked her interviewee two pointed questions in this line of questioning, "Did you have any other dream jobs you wanted when you were younger? What has been difficult about being a zookeeper?" These questions about obstacles, difficulties, and persistence were a central focus of the interview questions and demonstrated the girls understanding that accessing science careers could be difficult as well as their desire to know how to persist in these fields.

Some of the questions were very pragmatic and showed the girls interest in truly understanding the work their sponsors do as a way of understanding the realities of the work or potentially seeing themselves in the line of work. In her questions for an anthropologist Mariana questioned, "Could you give me a brief overview of what exactly you do in your field?" as a way of really understanding anthropology, its scope, and day-to-day work in the field. Mariana reported an interest in many scientific fields and really saw herself as a budding scientist. As such, she used the interview as a way to understand scientific work deeply and decide if the field was right for her. Leah asked the midwife, "How many births do you typically attend a month?" because she wanted to understand the work hours and expectations for a midwife. Laura asked the environmental scientist, "Can you tell me about your last major project?" Laura understood the project-based nature of her interviewee and wanted to know more about the types of projects the interviewee takes on. In these questions, the girls sought to understand the work of these women scientists and sponsors to better imagine themselves in the work.

Many of the students wrote interview questions that really tapped into the expertise of their sponsors, which is one of the greatest benefits of having a sponsor. Laura, who was deeply interested in sustainability wrote, "Do you believe there has been positive progress in society over sustainability? Why do you believe this?" This question invited the sponsor to share her deep knowledge and expertise and engage in a conversation with a young potential scientist who is very interested in sustainability. Bianca, who interviewed a forensic scientist asked questions about difficulties in forensic science but also asked, "What would you do differently to improve your work?" This question not only shows Bianca's interest in understanding the work of forensic science but also her desire to know about the ways that forensic scientists influence their labs and the quality of their work. Finally, Miranda who was paired with a local zookeeper wrote, "What is the most common misconception of your work?" Miranda's question illustrates her understanding that zookeeping is complex and her desire to know more. In this way, the girls tapped into the knowledge of their sponsors and included their own knowledge and understanding of these fields. In essence, through the interview question workshop, the girls learned skills for writing interview questions that provided them with access to fields they wanted to enter or understand.

Requesting an Interview—Writing Professional Emails

While many of the girls had prior experience writing emails in school, especially to teachers, very few had experience with what might be considered a professional email to an adult other than a teacher. In fact, as we both often find when working with incoming college students, many students believe that email is a genre for the past generation. These girls were no exception. Despite this generational divide regarding digital communication, nearly all of the sponsors asked that the girls contact them via email to either schedule an in-person interview or for an email interview. Furthermore, while we had hoped the girls would largely be conducting in-person interviews so that they could personally meet their sponsors, this research took place in a large metro area and the girls did not have transportation to meet the sponsors in person. Therefore, the email writing workshop became even more important since girls who wanted to participate in this larger network needed to send emails.

In the email writing workshop, we took a genre-based approach (Bazerman, Bonini, & Figueiredo, 2009) to composing emails. The genre-based approach invites students to learn how to examine the genre of a text, identify genre elements, and use the text as a mentor text for composing. Christina composed a mentor text for the workshop. At the beginning of the workshop, Christina and the girls talked about the mentor text and how this email was similar and/or different than others they may have sent. The girls immediately talked about the professional tone of the email and the language choices Christina used. The students identified the parts of the email, including: the subject line, greeting, moves made in the paragraphs, clear questions, and a signature line. As these genre elements were discussed, Christina marked them on her email and each of the girls marked their own (see Figure 4.3). Then using the mentor text as an example, the girls each began composing their emails. Some of the girls simply wrote to ask for an in-person interview, while many others sent along their interview questions for their sponsors to answer.

Many of the emails followed the model fairly closely since the girls were all essentially requesting an interview. However, we did notice some differences in a few key areas. The first is in connecting to the network. Many of the girls simply wrote, "I received your contact information from Dr. Christina Saidy and Dr. Jessica Early" since we made the majority of connections in the network. However, Aja, who connected with a psychologist in her family reflected this connection in her email by saying, "This is Aja, Lily's daughter, and I'm writing to you because I would like to interview you for a project at my school." Aja expanded the network by writing to a family member, and her email clearly shows that extension. Her email also shows a very clear understanding of the need to establish an ethos in a professional email. The girls who interviewed the science teacher at school reached out to that teacher in person. While they seemed relieved to interview someone familiar, they also realized they did not have a benefit of extending their own network into the professional realm. School was part of this network of women interested in advancing access to the scientists, and their sponsor was a familiar one.

In the emails some of the girls took the opportunity to share their sincere interest in science and the careers of the women they were interviewing. Bianca wrote to her sponsor, "It would be amazing if you could answer these questions." Mariana shared with her sponsor, "I've heard briefly about you and I'm very interested in your science career." Mariana had talked with us about her sponsor, a woman they knew well, and hoped to show that excitement and interest in her email. Laura reached out by sharing more about herself and why she was interested in her sponsor's career:

> During the summer of 2014 I went to a summer program at the The Honors College. During this time I took a class on sustainability, how it works, what could be done, ways it could be decreased, and why it is almost impossible. I have been interested in this field and what I could do to help in the future. When I went back to school my teacher asked

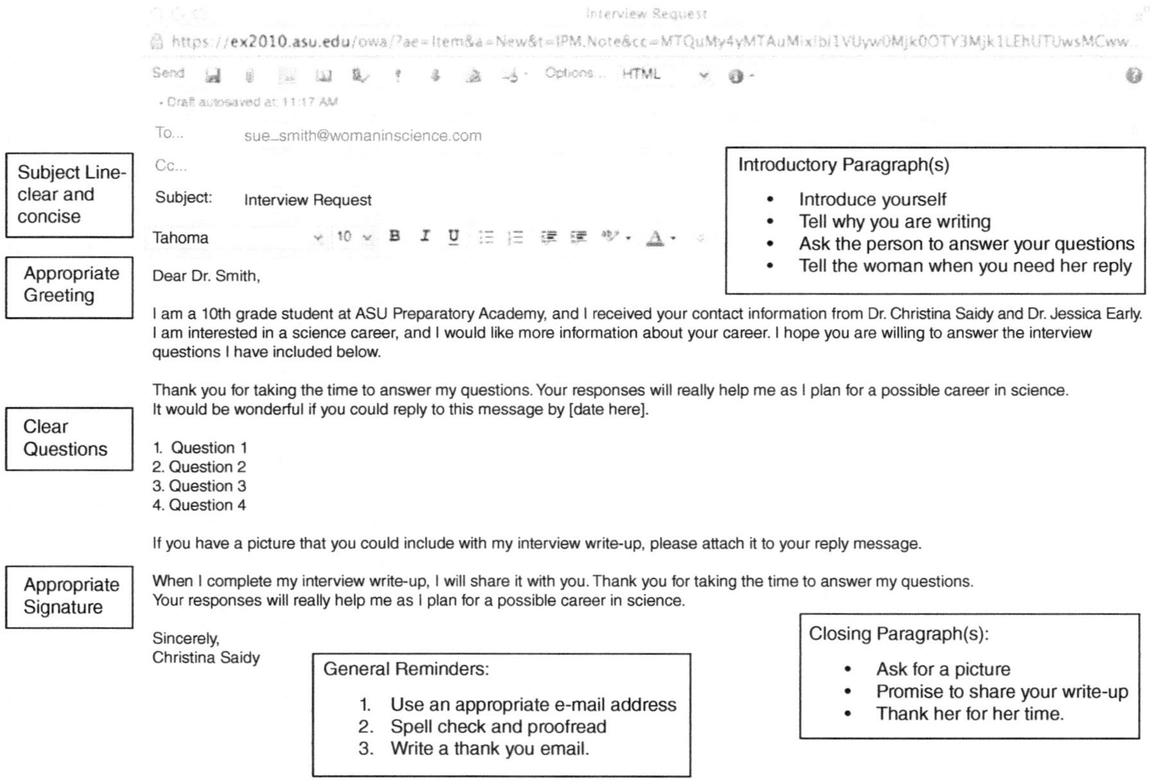

Figure 4.3 Email Mentor Text with Markings

> us to take an online survey [the university] had created. You would have to pick one of two photos about 70 times to see what they believe could be our future career. What they picked did not necessarily mean that we would have to be what they choose. At the end of my survey it gave me a choice to be involved in a future career in environmental studies, including this one. I would really appreciate it if you could sincerely and honestly tell me about your career. Further contact beyond this point would also be appreciated.

In her message, Laura clearly situated her interests in her sponsor's career field and expressed why her sponsor's experience could be of help to her. This clear articulation of interest in the email followed up by a request for future communication shows Laura's passion for her future field, need to know more, and understanding that she was carving out a spot in the network of women in this field.

At the conclusion of the email workshop, the girls sent their messages and waited excitedly and nervously for replies from their sponsors. Girls who were conducting in-person interviews prepared for their interviews. The mood in the room was a mix of excitement and nervous anticipation. The girls had taken a leap by reaching out to their sponsors, and they were excited to see how their sponsors would reply.

Sharing and Sponsorship—Replies From Interviewees

One week after sending their emails, the girls rushed into the GWS meeting. They came with interview notes and printed copies of email interviews. They were talking excitedly sharing pictures and things their interviewees said. Leah held up pictures she had printed of a midwife looking adoringly at a newborn baby. Mariana shared a picture of her interviewee—an anthropologist in a lab coat in her lab. Each of the girls had interesting facts they wanted to share.

As the girls excitedly shared what they learned, some things stood out to us and to the teachers. We had established this network of girls, teachers, researchers, and women scientists not really knowing how things would go. Our curriculum focused on developing literacy skills that would give the girls access to science, but once the girls wrote to their scientists, we were not sure what the replies would be. However, as we analyzed the interview notes and emails, we found that the women scientists served as positive sponsors providing information that would help the girls access science fields.

In all of the interviews, the sponsors replied to the girls with generosity and encouragement. In the anthropologist's response to Mariana, she wished Mariana luck in her education and career and offered to chat more if Mariana had more questions. The environmental scientist shared her work phone number and encouraged both of the girls who interviewed her to call if they wanted to know more. The zookeeper not only answered the interview questions in great detail but also provided some responses from a previous interview that she thought might be helpful. The forensic scientist wished Bianca good luck with making a career and college choice. In each of the interviews, the sponsors spoke openly and with an awareness that they were opening up doors for future women in science.

Many of the women shared with the girls the importance of finding a career they care about either personally or professionally. The pediatrician shared, "I knew as a young girl I wanted to be in a career where I could help others. My mom worked in the medical field as cytotechnologist which is a person who helps diagnose cervical cancer in women. I always admired how she helped so many women and I wanted to do the same." The psychologist shared a similar story, "I've always had an interest in understanding people, their actions, and why they act the way they do. Also helping others to grow as a person." This interest in following passions and helping others was evident in many of the interview responses and the girls were impacted by these responses since they had been following their passions or connecting to their lives in much the same way that the sponsors had. For example, Carola said she was interested in becoming a doctor, "because since I was little my mom used to have very bad migraines and would always get very sick and have to go to the doctor." The responses from the sponsors validated the ways that many of the girls were processing their career interests. The sponsors responded with connections to real life and to their passions, just as the girls had.

Many of the sponsors acknowledged the challenges of being both a woman and a scientist. For example, the midwife told Leah, "The biggest weakness of the profession for me is time. Midwifery is often a calling of the heart and can be so difficult to leave 'the job' at the office. It means time away from my family, missed birthdays, holidays, special events." The anthropologist told Mariana, "One thing that motivates me is the importance of being a role model as a scientist who is also a woman. I didn't have many examples of that growing up, and even in college and graduate school I wasn't sure it was possible to be a scientist and also be a parent and well-rounded person. I really love being a scientist, and a woman, and a parent, so I want to show others, especially girls and women, that it is possible." The anthropologist went on to speak about gender stereotypes in schooling and the need to prove she was as smart or smarter than the men in class. These sponsors showed the girls that many of the obstacles they had observed via their study of women in science actually did exist. However, they also showed that they have very successful careers they love despite the obstacles. These sponsors articulated an interest in sharing their struggles as a way to provide access to a younger generation of women scientists, like the girls in GWS.

As is often the case with sponsors, they highlighted the realities of the careers as well. For example, the zookeeper told Miranda that she loves her job, but Miranda should really consider the "working conditions and pay versus the rewards of the career." The environmental scientist, who also loves her career, echoed this concern about pay, telling Coco, "Get an engineering degree. Salaries are higher and the skills in engineering are highly transferable." The forensic scientist, who works for the police department addressed Bianca's question about how to improve her work by saying, "If there was one thing I could do to improve my work, it would be to hire more staff. As a governmental agency, we have very limited resources, and unfortunately we don't have enough staff to complete all of the work that is needed." In their reflections, the girls seemed caught off guard by the discussion of pay. Miranda was surprised that her interviewee "talked more about higher paying jobs." Laura, who interviewed the anthropologist, reported, "What surprised me is that it is not a high paying job." This is one of the important elements about sponsorship, that a sponsor often shares the "secrets" of a field, even if that means that they question their work or specialty. The environmental scientist said she may have chosen a different and higher paying career that was also in the sciences. The anthropologist shared with Laura that she loved the job despite the pay, a sentiment the zookeeper echoed. The sponsors provided the girls with factual and realistic information about the pros and cons of their careers as a way to support the girls in their own choice of career.

For many of the students, this experience represented the first time they had interviewed another person. The girls stepped outside their normal roles as students responding to teachers' requests and, in this interview, became the person shaping the conversation, asking the questions, and making sense of the information received. In the post-interview reflection, students reflected on the experience of being an interviewer for the first time and realizing what they could improve in the communicative and writing process. For example, Mariana wrote, "I wish I had asked more specific questions about her career and process of her work." Coco wrote a list of questions she wished she had asked her interviewee but had not thought of until after the fact: "How does she manage her time? What is her main focus in work? What benefits does she receive?" We invited students to email their scientists if they still had pressing questions and many of the scientists had also expressed an interest in an ongoing relationship with the girls in the future. In this way, the women scientists, the girls, the teachers, and the researchers truly became a part of a network. The girls learned to use their literacy skills access this network and to access their science careers.

Learning to Network: Genre Analysis and Mentor Texts as a Way In

One of the ways that writers and speakers show that they belong in a network is that they master the genre skills used in that network. In this chapter, we described the ways that the girls used email to access the network of women scientists who had agreed to be their sponsors. For the girls in this project, email seemed like an antiquated genre that was not relevant to their communicative practices in the real world. However, as they quickly learned, email is one of the most common genres in professional communication. If they wanted to access their sponsors, they needed to know how to access and reproduce this genre. This genre-based approach is one that we use in most of our writing classes.

To give the girls access to the information needed we began with a simple genre analysis of a mentor text. Here are some questions you can use to analyze a mentor text with your students:

1. Who is the audience?
2. In what context is this genre often used?
3. Who typically uses this genre and for what purpose?
4. How is this genre formatted (paragraphs, lists, greetings)?
5. What are any visual cues that are important (bullets, highlights, bolding, etc.)?
6. What is the level of formality in the language?

Learning to Network: Texts on Genre in the Classroom

Here are some resources to for incorporating genre-based instruction and using mentor texts in your classroom.

Books

Bazerman, C., Bonini, A., & Figueiredo, D. (Eds.). (2009). *Genre in a changing world.* Fort Collins, CO: Parlor Press and the WAC Clearinghouse.

Dean, D. (2008). *Genre theory: Teaching, writing, and being.* Urbana, IL: National Council of Teachers of English.

Gallagher, K. (2011). *Write like this: Teaching real-world writing through modeling & mentor texts.* Portland, ME: Stenhouse Publishers.

Online Resources

Purdue Owl Retrieved from https://owl.english.purdue.edu/owl/resource/1015/03/

5 Learning to Envision

> I thought it was really cool to write the profile essay to introduce my scientist but make it my own and carry it in my own way. I thought that was fun.
>
> —Mariana, tenth grade

Leah walked into the classroom the morning after receiving her interview response from her scientist via email. She exclaimed, "I can't believe my scientist actually wrote back to me! It was so cool to see the midwife's face and a photo she had taken at a birth!" Like Leah, other students expressed delight and surprise about the responses they received from their interviewees. The act of interviewing professional women in science represented a turning point in this project and propelled the work of our literacy community forward. The interviews made the project more than "work for school" and became something students were deeply invested in. The work became something the girls saw value in for themselves and for the literacy community they were a part of. They expressed to Jessica and Christina and to one another, how conducting these interviews "made the work feel more real."

The process of connecting to women in science through the interviews gave students the opportunity to experience writing as a social process, grounded in lived experiences, with real and applicable purposes (Bazerman, 2016; Early & DeCosta, 2012). Students held agency throughout this project, from the selection of the kind of scientist they wanted to interview, to writing interview questions, and initiating and conducting their interviews. Throughout, students were asked to use reading and writing to engage with and reflect on other people's lived experiences and professional choices as a means of thinking about and planning their own. The women scientists served as examples of possible life pathways students could envision, examine, and articulate in relation to their own dreams and interests (Singer & Hubbard, 2003).

Capturing the experiences of women in science through these interviews allowed the girls within our literacy community to see there is no one or no right pathway to becoming a scientist, to being a woman, to having a job, or to persisting toward an end goal. Many of the girls' initial understandings of what it means to work in the world in science were based on their familial, school, community, and cultural worlds (Moll et al., 1992). However, through the work of this literacy community, students expanded their understanding of what adults do in the world beyond the familiar. They learned from women in science who shared their stories of education, persistence, support, and hard work.

This chapter shares the interview profile component of the project, which was the final piece of the Girls Writing Science Project and took place once the girls had conducted their interviews. The written profiles gave students an opportunity to tell the story of the woman they had interviewed and to include their own lessons learned and perspectives about what it means to be a woman in science both for the woman they interviewed and for themselves in their own life pathways. For the profile contents, we asked students to introduce the woman they had interviewed, share her significant life and career pathways, and share what they had learned through the experience as young women and interviewers. This writing, which is described in depth below, focused mainly on the story of a woman in science. However, it also became a way for the girls to reflect on and envision their own lives, interests, and perspectives in relation to science.

Interview Profiles

Jessica and Christina often listen to National Public Radio's interview program *Fresh Air* with Terry Gross. Gross frequently interviews journalists from the *New York Times* or *The New Yorker* who write profile pieces based on in-depth interviews and contact with politicians, celebrities, academics, and other notable people. Profile pieces are a common genre form used in journalism to highlight the lives or contributions of individual people. An interview profile is a written genre featuring one person, in this case a woman scientist, and it usually goes into depth about significant events or details that helped shape the individual's story, achievements, and/or perspective.

In framing the interview profile assignment, we asked students to write an essay introducing the woman they had interviewed, sharing significant life and career pathways, and explaining what they, as young women, learned through the experience (see Figure 5.1). We thought the profile essay would be a perfect genre for students to become familiar and engage with the information collected through their interviews of women scientists. We planned four concrete mini-lessons to help give students guidance with the genre,

Interview Profile Assignment

We would like you to write a profile essay based on your interview with a woman scientist. This interview write-up will not only teach others about this particular scientist, but it will also share what this work has to offer others, and what you have learned or have you have grown through this process. You will share this piece with your classmates, teachers, and other young women you hope to inspire.

The write-up must include:

- A title
- A successful lead
- Describe background information about the woman scientist in an interesting way
- At least one interesting anecdote (story) about the scientist
- At least two quotes from your scientist
- A connection to your own life
- A successful conclusion—what you learned about yourself and who you want to be through this interview process

Figure 5.1 Interview Profile Assignment

which we detail next. These lessons included 1) writing a strong introduction, 2) embedding quotes from interview transcripts, 3) writing to reflect, and 4) revision.

Ways to Begin

The first mini-lesson we taught for the profile piece was on writing introductions. We gave the girls two options for starting their profiles. They could either begin with an anecdote about the scientist's life or a recurring theme found within the interview transcript. Jessica shared examples of these approaches to writing introductions (see Figure 5.2). She also provided an example from a profile piece in *The New Yorker* (White, 1989) about a marine botanist and biologist, Dr. Sylvia Earle, who was studying Prince William Sound, Alaska, after the Exxon oil spill. The author begins with a poignant anecdote of flying over the area of the spill looking out the window with Dr. Earle:

> Flying over Prince William Sound, Alaska—an expanse of silver-gray seas, bindingly white mountains, stark blue glaciers, wild islands, and rainbow-hued streamers of spilled petroleum snaking through thousands of square miles of ocean—Dr. Sylvia A. Earle looked out the window of a Coast Guard helicopter and shouted over the noise of rotors, "It's an inexcusable outrage!"

Tips on Writing Introductions

You will be telling the story of your woman scientist as a way to share about her work and educate those who might be interested. One strategy for writing an introduction for your interview write-up is to begin with a compelling story or anecdote to draw your reader in.

1. Starting with a compelling anecdote: Open with a small story that personalizes the essay topic. It may introduce who you are as a unique individual, pose the thesis or dilemma that controls your argument, or provide insight into your interests and values.
2. Starting with something thematic found within the interview transcripts: Open with a recurring theme or idea your interviewee talked about from her work or her life.

Figure 5.2 Tips on Writing Introductions

Next, Jessica modeled how students could begin their profile essays using a common theme or idea found in their interview transcripts or notes. The theme could be something their interviewee mentioned intentionally throughout the interview, or an idea that emerged from the transcript notes. Aja took up this approach in her write-up of a psychologist:

> [The psychologist] faced many problems living away from her family at a young age. She faced the fear of being alone all by herself and the fact that the school she attended was far from the place she lived. She also faced economic problems like paying for her education, housing, and personal needs all by herself. Although she faced all of these obstacles, nothing snatched her passion to help others.

Instead of listing facts in chronological order or simply restating or summarizing the transcripts and interview notes, students could begin by telling the story of their woman scientist in compelling and effective ways. Anita introduced her interviewee, a pediatrician, in her first paragraph and shared how she was inspired daily as a young girl by her mom:

> Her mom was a cytotechnologist who looks through microscopes to diagnose cervical cancer. She knew that her mom was doing important things in the world. When she was little, she was proud of her mom and knew that her mom was passionate about her career. Following in her footsteps, she wanted to help people and loved to learn about the human body. Once she was in high school, she knew she wanted to be a physician.

Other students, like Destiny, began the interview profile sharing what she learned from her interviewee about the process of being or becoming a scientist. Destiny's scientist was a science teacher working toward her medical degree. "I realized the road in becoming a doctor can be a long and difficult process. Ms. C is a high school teacher and is going to school to become a family medicine doctor. She is a person that is experiencing this process in becoming a doctor." Ms. C says, "becoming a doctor gives me a chance to combine helping people and science together."

We circled the room while the girls were writing and met with them one-on-one to give feedback to help them revise their introductions to effectively frame their essays. Leah chose to begin her essay by sharing her scientist's love for midwifery. "[The midwife] was originally planning to follow her father's footsteps and become a paramedic. Through her exploration of the medical field, she realized she loved the idea of becoming a midwife. She remembers feeling terrified of her first delivery. But, she was thrilled after it was over. She found a way to follow her heart and do what she loves."

Teaching students different strategies for beginning their essays provided structure and choice as students began writing. Many of the girls in our group considered themselves reluctant or hesitant writers prior to the project as reflected on self-efficacy Likert-scale survey we created. We derived the scale from Bandura (1986) and Shell, Colvin, and Bruning (1995) and Early and DeCosta-Smith (2011). This questionnaire consisted of fifteen questions connected to facets of writing associated with writing about science and the profile essay genre in a response format. The ten-point scale ranged from "not at all confident" to "completely confident" with 1 being the least confident and 10 being the most. We distributed the survey on the first day of the workshop prior to any instruction and then again at the completion of the project. In the beginning, the average score of writing confidence associated with writing about an interview the student had conducted was 5. After the project, this number increased to 8. At the beginning of the project, the average score for writing introductions for an essay was 5 and after the workshop it rose to 7. When we interviewed students after completion of the workshop, many pointed to the introduction workshop as one of the most helpful pieces of the project in helping them move forward as writers. For example, Clara explained, "The introduction lesson was so helpful because you taught us certain things about writing an interview and you gave us examples of interview profiles. I loved getting a chance to learn how to write a successful lead using an anecdote or theme. It helped me get started."

Embedding Interview Quotes

One of our goals for the writing of the profile pieces was for students to learn to tell the story of another person using information they had gathered themselves through the interview to tell the story. To do this, we taught the girls to weave interview quotes into their profile pieces and not provide raw transcript without context and explanation. First, we provided model texts to show examples of other writers doing this work in their writing. Whenever we brought in model texts to teach specific writing strategies, we made it a point to bring in texts that included content on women in science. As we were teaching writing, we were also working to deeply immerse the girls in examples of women in science and their career pathways.

The mini-lessons were about teaching writing strategies and setting students up with points of entry into their essays. Many of the girls considered themselves struggling writers and we realized the thought of a blank page and a big writing task ahead of them would be anxiety inducing. Providing model texts, strategies, and choices for writing the introduction to the profile piece gave students a way in to the essay and strategies the girls could take and use in other writing assignments and tasks in school and beyond. Mariana shared her appreciation for this lesson in her final interview after completing the project, "I will definitely use the intro writing ideas we learned in the workshop. I really liked that. It gave me a different sense to writing and to style."

In the skill-lesson on weaving interview transcript quotes into the profile essays, we began by showing the students a TED talk presented by Lina Colucci, a biomedical engineer (Colucci, 2014). Colucci is known for her work in the United States to develop a portable sensor for congestive heart failure patients. She is also known or her early work, as a high school student, to design and engineer a more comfortable and lasting ballet shoes. Colucci became interested in science and engineering from a young age alongside her passion for dance. In Colucci's TEDx talk, she describes her work as a scientist and as a dancer and how she followed her interests and made connections between them. After watching Colucci explain her own pathway and work, we read an interview of Colucci from *Science*, a popular science magazine (Pain, 2015). The piece is a transcript of a written interview exchange with the author. Christina read the piece aloud, so students could see how interview transcripts are sometimes published as transcripts alone without the author doing the work of profiling, contextualizing, or describing to share the story of the person. Christina used this as a model of an interview genre as a genre by itself, but also to point out that the profile we were asking students to write was something different. It was more extensive and included commentary, story, and reflection. Next, we read an article about Calucci's work as a woman in science that profiles her and her work as a scientist (Duke University, 2014). Within this profile article, we asked students to highlight direct quotes the author embedded into the piece based on interviews she conducted with Colucci. For example, the author used the following quote to give a sense of Colucci's pathway as a scientist in college, "I just wanted to design cool stuff. Studying mechanical engineering at Duke was a natural choice. However, throughout most of college, I had no definition of 'cool stuff.' My friends and I spent a lot of time discussing what we wanted to do with our lives and I was struggling with how to manage 'cool' and beneficial to humanity" (p. 45).

We created a handout to show the girls basic strategies for using direct quotes within a longer piece of writing using quotes from Pain's article (2015) (see Figure 5.3). As the girls began writing their first drafts of their profile pieces, they practiced supporting their ideas with evidence from the interview transcripts and our help. For example, Leah used a direct quote from her interview and set it up to read smoothly, "Before becoming a midwife, [she] was afraid that she wouldn't find employment as a Certified Nurse Midwife, 'The nearest CNMs were over an hour away and the area I lived was extremely rural and did not have any clinics or places I could find work easily.'" Daniela used a quote from her interviewee to describe her inspiration. "Ms. C has one major inspiration in her life. When she was a little girl, Ms. C's pediatrician had a big impact on her and served as her inspiration. She explained, 'As I grew closer to my pediatrician, she encouraged me to become a doctor.'" Some of the girls in the group really struggled with this part of the writing process. For example, Coco only included a single quote from her interview in her final profile essay. We realize looking back, that this is a place we could have used more time with the girls to support their writing. While we introduced these skills, provided model texts, gave concrete tips, and offered feedback, we realize we did not have access to the students long enough to give them the support they needed to completely succeed with this writing invitation. Taking direct quotes from a transcript and embedding them into a text is a difficult writing task.

The Power of Being Read

After the girls completed first drafts of their final profile essays, we asked them to turn in their drafts, so we could provide feedback and suggestions to help them shape the pieces. Student's first drafts of their profile pieces revealed that they could detail the stories of the scientists on a basic level; however, even with the workshop on introductions and embedding quotes, we wanted them to take part in peer review to gain experience reading the writing of their classmates (Early & Saidy, 2014a). Jessica and Christina also gave the girls substantial written feedback on yellow sticky notes, which they placed on their first drafts to help guide their next steps.

Revision, a critical aspect of writing effectively, is one of the many small steps writers take part in as they produce and polish texts (Harklau & Pinnow, 2009). One of our goals in this workshop was to provide students with

Tips on Embedding Interview Quotes

Quotes do not work when they are plunked in—you need to prepare your reader for the quotes you are going to use. Here are some ways to weave your quotes into your writing effectively:

1. Assume your reader knows nothing about your person or subject—provide the right context/background for the quote.

Colucci, a ballerina since she was five, describes how ballet shoes haven't changed in the past two hundred years. She says, "They are still extremely primitive." As an engineer, she decided this needed to change.

2. Your quotes should be short and sweet (see above).

3. If you find a perfect quote to illustrate your point, but it is too long, then there is a way to shorten it without changing the author's words. You can leave three dots . . . to indicate where you have left out words (see bold example below). This is called an ellipsis.

Whole quote: "*In all these endeavors, a final product is born from feelings and fragmented ideas. Also, I think everyone needs something that feeds their mind, body, and spirit. Of course, most activities don't serve just one purpose. Dancing on stage is a spiritual experience as much as it is a physical one. Learning or choreographing new steps can be as intellectually stimulating as doing lab research.*"

Shortened Quote: "*I think everyone needs something that feeds their mind, body, and spirit. . . . Dancing on stage is a spiritual experience as much as it is a physical one.*"

4. Remember—your words should carry the piece—NOT your quotes! Quotes are a way to put the person's voice along with your words. It is a way to use the interviewee's words to enhance the story you are telling.

Colucci is a classically trained ballet dancer and an accomplished clarinet and saxophone player. She has been combing arts with engineering with arts since an early age. For example, she started a multi-year project redesigning traditional ballet pointe shoes while she was still in high school; the project yielded her first publication. In 2011, she did a summer internship at Nike's research labs in Beaverton, Oregon to learn more about shoe design. For her, arts and engineering are inseparable, because she believes "*everyone needs something that feeds their mind, body, and spirit.*"

5. Have fun writing and using quotes to tell your scientist's story.

Figure 5.3 Tips on Embedding Interview Quotes

Figure 5.4 Jessica offers students feedback on their profile assignments

Source: Charlie Leight

an opportunity to experience substantive revision as a part of the writing process. Substantive revision can help improve the overall quality of a piece of writing, improve the organization and presentation of ideas, and strengthen a line of argument (Bridwell, 1980).

Here is an example of Josephina's first draft of her profile essay before she received any peer or teacher feedback:

> Mrs. C is the only one in her family that understands science and is going into the medical field. Her biggest challenge in school is studying for the tests and studying for them. But she has her heart on her career well that's what is important in becoming a primary doctor.
>
> What I took out of this interview that you need your heart to be set on your goal in order to accomplish it. Hard work pays off at the end because if you are set for what's expected you are ready for any life challenges you get. I learned that in becoming a doctor you are willing to help out not only one person but the whole community. The good thing about helping people is that . . .

Jessica and Christina divided up the job of reading and providing feedback on the first drafts. Christina read Josephina's draft and gave her the following two tips:

1. Tell a little bit more here about what you learned about Mrs. C and her plans for medical school.

And, in response to the second paragraph:

2. This is a nice connecting paragraph. Try to finish it in a strong way connecting to Mrs. C. Share more of your response to this interview and what you gained from it.

In her final draft, Josephina more than tripled the length of her piece. She added more detail about Mrs. C's schooling, goals, life choices, and inspiration. She wrote how Mrs. C was inspired by her own children's pediatrician "because this doctor was important in her own life as part of a family and as a mom." At the end of her profile, Josephina took up Christina's advice and added in what she learned from the interview with Mrs. C and what she will take away from it:

> I want to be like Mrs. C because she wants to help people and so do I. I also want to be like Mrs. C and do something with science. Mrs. C is the only one in her family that went into the medical field and I will be the first person in my family to do the same. We are both scared of losing a patient because this is something hard to accept. What I took away from this interview as that you need your heart to be set on your goal to accomplish it. Hard work pays off at the end because if you are set for what's expected you are ready for any life challenges that come your way. I learned that in becoming a doctor you are willing to help out not only one person but a whole community.

The above selection from Josephina's profile serves as an example of how students used each of the steps of the writing workshop to fill out their essays, take up the feedback, reflect, and build toward a complete final profile.

Some students, like Anita, struggled writing the first draft. Anita's first draft was only representative of the writing in the introduction and embedding quotes workshops and she had not extended the piece further. Jessica responded with the following ideas and questions to help her think about ways of expanding the essay:

1. Maybe talk a little bit more here about the challenges and rewards of being a pediatrician.
2. Give a little more detail here. What did you learn specifically? Why did this make you more passionate? I like how you talked about yourself. Can you connect to your interview?

We tried to provide the girls feedback made up of open-ended questions to help them see their piece from different angles and to write in more depth.

In the next draft, Anita answered these questions and expanded her piece dramatically. Along with her introduction she had added:

> Once [the pediatrician] finished school, she became a pediatrician. She would see patients every day and help them with a situation. One day, a mom came with her son and asked her to find out why her son was very fussy. [The pediatrician] said that there is nothing wrong with him, but the mother refused to believe that and came back a couple of times. [the pediatrician] kept saying there was nothing wrong. As months passed, [the pediatrician] got some flowers and box of candy, then she realized that it was from the mother who refused to believe that her son didn't have something wrong with him. It said, "Thank you for being patient with me and my son." [The pediatrician] was very grateful and knew she was doing something good.
>
> I will take away from this interview that, at times, it can be very stressful to be a doctor. But, overall, if it is something you are passionate about, then you should go for it. It's beautiful to see others appreciate what you do daily for them and to see how a passion for medicine is developed over time. I like the idea of knowing you can help someone and they can get better and you can get love from them.

Participation in the revision workshop not only changed writing, but also expanded students' understanding of what it means to be read, to receive feedback, and to revise written work.

Final Essays

The community of writers shared their final profile essays aloud with their peers on the last day of our workshop and sent their write-ups to the women they had interviewed as a culminating written exchange. As the girls finished their revision work and their profile essays were ready to be shared publicly, they took turns in a chair at the front of the room and introduced the class to their women in science by reading their essays aloud. The following is Aja's final profile piece about a psychologist, which she proudly read to her peers at our final reading celebration:

> *Everything is There for a Reason*
>
> [The psychologist] faced many problems in her path to becoming a psychologist. Among these problems were being away from her family and facing the fear of being alone. The school she attended was far from the

> place she lived, and she had financial problems, such as paying for her education, housing, and personal needs. Although she faced these obstacles, nothing snatched her passion for helping others.
>
> [The psychologist] is a psychologist. She is interested in other people's behavior and why they act the ways they do. Her interests moved on to helping people to overcome their problems and grow even better as people. She wants to help people learn to be happy with themselves. She says, "It gives me a great feeling seeing people better than how they were the first time." Psychologists are people who listen to you. They help you overcome your problems and face difficult situations.
>
> Like most people, [the psychologist] has second-guessed her career. "If I had the opportunity to change my career, I would choose medicine," she said. Her biggest fear as a psychologist is not being able to help one of her patients. Her goal is to help people face their problems and change their attitudes about obstacles they face.
>
> From this interview experience, I take away the feeling that helping others is the best thing to do always. You never know if you are the only person helping someone else. In the end, you might help them to feel happier or to live better with themselves. This is one of my future goals. I was once there for someone who was contemplating suicide. This was the first time I realized that helping people is what I want to do. The person I helped had a lot of problems, but she told me I helped by being there for her. This interview made me grow as a person and my passion for my future career as a psychologist increased.

Aja's profile piece not only teaches others about the work of a remarkable scientist, but also shares a sincere connection to this woman's work. The profile piece is a reminder of the way learning about other people's life stories provides opportunities to envision and "try on" new ways of thinking about and acting in the world. Through her access to, interview with, and write-up of an interview with a woman scientist, she began to form her own understanding of what it means to be a woman in science and, more specifically, a psychologist.

Pathways of Women Scientists

The interviews gave students a window into the diverse pathways individuals follow as they prepare for, enter into, and participate in science-related careers. As part of their interview profile essays, students reflected on the interview experience and on the workshop as a whole. They emphasized how this writing project represented a new kind of learning, requiring them to reach beyond the classroom and out into the community to understand the intricate choices, pathways, and actions that women take to become successful scientists. After students completed the writing of the main part of their profile piece, we asked them to write a reflection about what they learned through the experience of interviewing. We asked students to write responses to the following questions to help guide this final piece of the profile essay:

1. Please take time to share what you learned in the process of writing your profile.
2. What did you notice about your writing process?
3. What did you learn about the scientist you studied that you will remember?

In her concluding paragraph of her interview profile, Leah wrote, "I found out you learn every day in a science career." Bianca appreciated learning specific writing skills, "I will take away knowledge and skills from this interview." By extending their learning networks beyond the classrooms and taking part in the interview process, students articulated how they learned how to rely on themselves in ways they had not before. Some of the skills are learning new forms of writing and being more open minded with my thinking and writing. Josephine's biggest take away from the project was the importance of setting her mind to something and working toward it:

> What I took out of this interview project was that you need your heart to be set on your goal to accomplish it. Hard work pays off in the end. After you learn what is expected to accomplish a goal, then you need to set yourself up to accomplish it.

For many of the students, the interview process expanded what they had thought was possible for them as young women. Coco wrote,

> I never really put to mind that women can balance a job like Julie's and take care of a family. I appreciate women like her and all they do. After weeks of constantly coming to the workshop, I got to see more of why I enjoy science. I understand what I have to do to get where I want to be and make a mark for myself. I got to see women do things I didn't even know were possible. My inner feminist came to the surface and I saw women can be as smart and as equal to men.

Anita wrote, "The most surprising thing I learned from my scientist is sometimes she is so busy she doesn't eat lunch because she needs to get all of her patients. Also, she had eleven years of training." Leah interviewed a midwife and it was surprising for her to learn this is not a high paying job and she wondered why. "I don't understand how midwives make so little money. Aren't they doctors?" We talked about the differences in pay in medical fields and how some of these differences have historical patterns connected to gender roles and gender bias. We also talked about how these roles continue to change as women have entered the workplace and received medical degrees and training (e.g. historically, men were doctors and women were nurses and midwives are predominantly women).

Other students learned how women scientists juggle their work responsibilities with their home lives. Mariana wrote, “I learned that she [my scientist] is a mother and she also teaches.” Anita reflected on the busy life and workplace of the doctor she interviewed, “I learned that she is a very busy person. She only has 15 minutes for each patient she sees, and she also has to balance out her personal life and work.” Coco’s interview gave her insight into how some scientific careers provide more job security than others. Her interviewee’s career is entirely dependent upon grant writing. Some students were truly surprised to learn how much the scientist they interviewed enjoyed her work. For example, Clara wrote, “The person I interviewed was really into her job. She didn’t say a single thing that was negative about her work. This was remarkable to me!”

The girls had different reactions to the life stories of the scientists they studied. In each case, they expressed how the stories of the women touched them in new and unexpected ways. Clara wrote in the conclusion of her profile piece about a pediatrician, “The interview inspired me because a woman made it to college and actually has a career doing what she wanted in life. This gave me actual details about how college will be and how difficult it is. Knowing the pediatrician made it and became a pediatrician lets me know that I also can make it.” Clara wrote in her profile piece how she was drawn to her scientist because of the similarities between the archeologist’s interests and her own and because she felt she made a connection with the archeologist as a person. “This interview meant a lot to me because I enjoy digging and looking at things I have found. I enjoyed interviewing [the archeologist] because we related. I will take away from the interview that it is okay to get a little messy and that this is just part of research. I’m glad that I found everything she had to say really interesting.”

What It Means to Envision

For many of the girls in our group, the interview process expanded what they had thought was possible for them as young women, “I never really put to mind that women can balance a job like Julie’s and take care of a family. I appreciate women like her and all they do.” Coco wrote, “After weeks of constantly coming to the workshop, I got to see more of why I enjoy science. I understand what I have to do to get where I want to be and make a mark for myself. I got to see women do things I didn’t even know were possible. My inner feminist came to the surface and I saw women can be as smart and as equal to men.”

Within the sociocultural framework of this project, we examined the ways the GWS project supported and worked to extend literacy funds of knowledge for students. We define literacy funds of knowledge as literacy skills individuals need to participate effectively, efficiently, and with success in college, the workplace, and the community (Compton-Lilly, 2009; Early & DeCosta, 2012; Moll et al., 1992). Through the interview responses, our literacy community collectively experienced the way life pathways are not straight lines or perfectly determined. These pathways can be unexpected, interrupted, frustrating, fascinating, and full of adventure. The girls in this group learned how science careers are shaped by choices to persist, to follow inspiration, to listen, to have families, to have security, and to achieve happiness. In school, so often, students are asked to think about the future in one direction only—forward. This writing project gave the girls an opportunity to think about the future in a different direction—from the career backward. We designed and implemented this community of practice as a place to build social and cultural literacy capital within a school-sponsored space.

Finally, the girls experienced what it felt like to reach out to and receive responses from people in their community as a community. As the girls received responses, they compared their interview experiences with those of their peers. They shared the stories of their women in science along with the obstacles they faced in the interview process and ideas for their final profile pieces. Writing to reach out to, communicate with, and make sense of the experiences of women in science, gave the girls within this literacy community a common denominator with which to work, think, share, and act. Within this community, students bounced ideas, obstacles, successes, and questions off of one another and off of us. In this way, we formed a literacy community of practice with reading and writing at the center of our shared work together.

Learning to Envision: Interview Project Steps at a Glance

1. Students read, watch and discuss interview examples.
2. Students prepare and participate in a practice peer interview in class.
3. Students receive Interview Project requirements.
4. Students participate in an in-class workshop on choosing an interviewee and preparing interview questions.
5. Students arrange and conduct interviews during class time or on their own.

6. Students write a First Thoughts Letter in class.
7. Students read examples of interview introductions and then write their own.
8. Students receive Interview Writing Instructions and proceed with writing.
9. Teach mini-lessons on different genre elements for writing profile essays.
10. Students peer review profile essays and turn in to receive teacher feedback
11. Students revise using feedback from peers and teacher.
12. Students write a Final Reflection letter and turn in their interview write-ups for revision suggestions.

Learning to Envision: More to Explore

1. National Writing Project Blog Talk Radio. (2016). Intersections: Powering Science Learning Through Partnerships.
2. NSTA (2016). Outstanding Science Trade Books for Grades K-12: *www.nsta.org/publications/ostb/*
3. Carroll, B., & Baker, T. (2016). Working at the Intersections of Formal and Informal Science and Literacy Education: *http://csl.nsta.org/2016/03/working-at-the-intersections/*
4. Elliot, L. A., Jaxon, K., & Salter, I. (2016). *Composing Science: A Facilitator's Guide to Writing in the Science Classroom*. New York, NY: Teachers College Press.

6 Conclusion and Call to Action

> It was amazing getting to know and experience the way these women are scientists and have gotten so far in their lives. What surprises me is they have so much to deal with but they still do because they love it.
>
> —Laura, tenth grade

We are deeply committed to working toward social justice in literacy education and research. One of the ways we believe teachers and researchers can work toward equity in schools is through the creation of communities of practice (Wenger, 2006) using literacy to explore disciplines, careers, and opportunities beyond classroom walls. Through this kind of community, students may use reading and writing to dream, ask, network, and envision possibilities for their future lives. Moreover, this project specifically values literacy as a point of access into disciplines too often made inaccessible for Latina girls.

In this final chapter, we return to the original tenets that guided this work and reflect on what we have learned through this process as literacy scholars, teacher educators, and women:

1. This was a feminist project,
2. This project was about equity and access, and
3. This project was about creating a literacy community.

As women scholars and teacher educators, we deliberately created a space within a school for girls, supported by women, as a way to access successful women in science. We did this as feminist and social justice acts. Our intention was to give Latina girls, a population too often underserved and underrepresented in the sciences and in advanced literacy courses, a chance to access other women who have overcome barriers to success in a male-dominated fields.

We did this in a state where Latino students and families have been systemically targeted, and students have been sorted and tracked away from opportunities like this one. We worked to create a learning space and curriculum where Latina girls were immersed in rich literacy experiences to access future possibilities in the sciences for them as young women. We did this by creating a new school-based literacy community of practice where girls could build on and add to their literacy knowledge base. The girls also built on their relationships with one another, the classroom teachers, the researchers, and the women scientists. They took part in a network of women helping women. They witnessed the work the researchers and teachers put in to the project and they, in turn, supported one another and learned to receive support from unfamiliar sponsors. Because we see literacy as a social practice grounded in context and purpose (Prior, 2006), we created a space for the girls to dream, ask, network, and envision possibilities for themselves in science through access, exchange, and support from others.

As we came to the conclusion of this research project, we wanted to know the impact of the project on the girls as writers, students, and potential future scientists. We did this in two ways: 1) we asked them to return to the drawing of a scientist that they did on day one and, 2) we asked them to complete a closing survey to reflect on the project.

Drawing of a Scientist

At the end of GWS we returned to the drawing a scientist activity the girls did on day one. We provided them with markers, crayons, and papers. At the top of the paper, the instructions read, "Think of what a scientist looks like and then take a few minutes and draw the scientist you imagine on this paper." They then turned the paper over and the instructions read, "Now that you've drawn a scientist, take some time and write down what a scientist looks like and what a scientist does."

In analyzing the images, we noticed how the girls used images to envision women in science based on their notions and understandings of scientific work and fields. In our analysis, we were struck by how all of the scientists pictured were either gendered as women or gender neutral. The pictures showed an expanded concept of scientific work from the images the girls had created at the start of the project. For example, Mariana drew a picture of a woman in regular clothes with two thought bubbles over her head (see Figure 6.1). In one thought bubble it says, "There's a problem" and in the other it says, "How can I fix it? Hmm . . .?" On the back Mariana wrote scientists, "think deeply about things and how they can help solve problems." Mariana's

Figure 6.1 Mariana's final drawing of a scientist

conception of a scientist as a problem solver shows a complexity of her own thoughts about science. To her, science does not have a specific place and is not defined by the type of work. Instead, science is defined by its connection to thought and complexity. Josephina drew a non-gendered person with thought bubbles containing symbols of science (i.e. a beaker) and a pencil. Although her picture shows traditional scientific symbols, Josephina wrote, "A scientist needs to be focused and be ready to keep learning science because science is expanding. A scientist helps the world, people, and community." Josephina understands that science changes and scientists are perpetual learners whose work benefits people. Daniela drew two pictures. One is a woman with dark hair in regular clothes and the other is a woman with dark curly hair in a lab coat. Destiny wrote, "A scientist could have a lab coat or regular clothes. A scientist is many different things. She can work with chemicals, animals, plants, germs, shoes, or anything really. She can work at a lab or at a school or from her house." Daniela's writing shows that she and the other girls began to see that a scientist's work is not limited to a lab or traditional setting and scientists are often just regular women who do good work.

Laura further developed the idea of a scientist's place and work. She drew a picture of a woman under a tree with a magnifying glass pointed down at the ground. The woman is intensely studying the ground under the tree (see Figure 6.2). In this picture, the scientist is studying the natural environment to which everybody has access. On the back of the paper Laura wrote, "A scientist can be anyone and can do a variety of things. A scientist looks dedicated and encouraged by the public to continue the good work they're doing. A scientist looks like a person who is confident with the facts they hold." Laura's reflection and picture illustrate her understanding that science takes place in the public realm and that it is important to the public. Laura's understanding that a scientist can be "anyone" suggests that she can also be a scientist if she works with facts, or scientific knowledge, in the public realm. Estrella furthered the idea that science is available to all. She drew a pair of eyes with glasses and wrote above it "a scientist can look like you and me." On the back of her paper, Leah wrote, "You can be a scientist, I can be a scientist." One of the primary things we noticed in looking at the final scientist drawings is that the girls came to believe that science is accessible to them and to others. While they know that scientists have a specific skill set, they also believed they could learn that skill set and participate in public life as scientists. Through their participation in the GWS project, the girls came to envision scientific work as work within their reach.

Closing Survey

On the last day of the workshop, we asked the girls to fill out a closing survey, which included six questions to reflect on their experience in the literacy community. The questions included the following topics: 1) what they learned from the workshop on women and science, 2) what they learned about writing interviews, 3) what they learned from exposure to a variety of literacy resources on women and science, 4) what they learned about writing,

Figure 6.2 Laura's final drawing of a scientist

5) what writing mini-lessons were most helpful, and, finally, 6) what they learned about themselves.

In response to the first question, the girls described how they learned what it means "to become" and what means "to be" a woman in science. The process of becoming a woman in science was eye opening for many of the girls in our community. For many, they realized that becoming a scientist takes perseverance and hard work. Estrella wrote, "I learned that all of these women tried their hardest and finally achieved their goals in a science career." Laura wrote, "I've learned that everything in the science field is not very easy to accomplish. You have to work for what you truly want. You wait years and years, but you will get it one day. You have to be careful about what you truly want to be." Clara shared what she learned about her goal of becoming a doctor by gaining admission to medical school. "I will remember from this project that getting into medical school will be difficult and maybe even take up all my time, but I still want to proceed with this career into the medical field. Getting a response from this interview and reading it inspired me to work even harder and also have even more of a passion for medicine." Carla wrote about her new awareness of gender inequities that exist in the sciences, "I've learned that a lot of the women science don't get recognized for what they do."

The girls also wrote about what they learned about what it means to be a woman in science. They reflected on what it means to take part in the daily work of a scientist, to be a part of a larger field, and to be a person juggling multiple roles. Clara realized how women in science often face gender inequities: "I've learned that a lot of women in science don't get recognized for what they do in the same way that men in science do." Coco shared, "Women do not fully get the recognition they deserve and have to work so much harder to get ahead." Angela was overwhelmed and inspired by the project and what she gained: "I learned that what women in science do is incredible and just awesome. What they do daily makes a huge difference in the world. The person I interviewed saves lives daily for children. How cool is that?!" Cynthia was equally inspired, "This was so amazing! I learned so many things about what and who I want to be when I grow up. I am so grateful for this opportunity."

The students wrote about the variety of resources we provided about women in science. Some students, like Carla, shared how they appreciated the way the various texts served as models for her own writing: "The resources were helpful because they gave examples of certain interviews or what women in science do." Mariana appreciated how the different kinds of texts and resources provided multiple ways of seeing and understanding the work of women in science. The texts helped the girls to ask questions that opened up this way of seeing, "These texts and online resources exposed me to many different studies and careers that are interesting and caught my attention to do myself that I didn't know about before." Other students, like Clara, found the examples inspiring. They helped her to dream into being and believe that it really is possible to be a scientist and achieve a challenging life goal. She wrote, "It was helpful to see all of those women succeed in science and it was inspiring to see them reach their dreams." The girls also reflected on how their exposure to different kinds of science careers and pathways made them see there is no one or right way to become or to be a scientist. Leah shared, "There are a variety of ways you can get into science and not only by action." And Angela wrote, "It helped me understand that there are so many fields in science and every woman does something amazing!" In her survey response, Aja shared how she was uniquely aware of what we were trying to do by exposing her, and the literacy community as a whole, to so many examples of women in science: "It is so helpful to introduce science to young people and adolescent girls because less than 30% of women choose a field in science. People underestimate girls and then girls underestimate themselves. These examples are showing us we can be smarter and better."

One of the things that enabled the girls to dream and envision future possibilities for themselves in connection to science was the further development of their literacy skills. The project pushed the girls, as writers and readers, in new ways. In the survey responses the girls made it clear that they found the writing they did for this project quite challenging. Some of the girls commented on learning new genres including professional emails, interview protocols, and interview profile essays. For example, Mariana wrote, "I learned that there is a different format to writing up an interview." Other students shared how it felt new to them to focus on one topic and format of writing for so long. They were used to writing only short pieces for their classes and moving on to new topics. Students also realized that the quality of their interview questions led to the depth and breadth they could write from in the interview profile essays. Their interview questions and responses became the data with which they could write-up their profiles and if the interview questions were weak and they got weak responses, they then had little to work with in their profiles. Josephina reflected, "I learned that there are questions you need to ask in order to get strong answers in an interview. The challenging part was interviewing someone because I was shy and everything was interesting to me. I realize I should have been more focused with my interview questions." Clara also reflected on her writing process and shared how she felt stuck and overwhelmed beginning some of the writing assignments. Then, she surprised herself by rising to the challenge: "It was challenging writing a minimum number of interview questions for my interview protocol. Yet, I wanted to ask a million. Once I actually sat down and started writing, my ideas began to flow." Tina felt a responsibility to write her interview profile well and with care and attention to the representation of her woman scientist. She wanted "people to see how amazing she was and as I saw her." The girls took on a great deal of responsibility in the write-ups of the interviews. They wanted to represent the women scientists they had interviewed fairly and well. They felt connected to these women as their sponsors and invested in their work. The surveys confirmed the ways this work propelled the girls to dream, ask, network, and envision their possible lives in connection to science.

Limitations

As in any research, this study has certain limitations that should be noted and addressed for future projects. One limitation is the relatively small number of participants in our group. Future studies could include participants across multiple classrooms or school settings. Due to the limited number of participants, these findings cannot be generalized to represent all first-generation Latina high school writers; however, the descriptions of these students provide a rich account of their experiences and mark an important starting point for understanding ways of supporting Latina girls' through the design and implementation of literacy communities of practice within school settings. A second limitation of this study was the relatively short time-period with which we conducted the project. With more time, we could have gone into more depth in our teaching and the girls would have had more time to focus on writing, revising, and presenting their final profile pieces. Although the brevity of this project was a limitation, it can also be interpreted as one of its strengths. The relatively short duration of the workshop makes it is accessible for researchers and teachers to create similar literacy communities within their school settings. Future studies could extend our findings to understand how this project may be expanded and modified to fit other academic disciplines beyond science and support the needs of diverse student populations across grade levels. Future projects could also do more to examine the impact of participation in this kind of community of practice on girls' educational and career choices and pathways. We did not follow up with the girls after completion of the project and we realize this would have been fruitful.

Looking back on the project and thinking of ways to strengthen it for future years, we discussed ways to include the women scientist sponsors and students' parents. In future projects, it would be powerful to give the girls a chance to meet their women science interviewees in person and to invite the sponsors to the final sharing of the profile writing. Furthermore, we imagine parents would have been impressed with their children and supportive of their efforts. At the secondary level, parents are often not as involved or connected to school-based learning and communities because of the students' age and independence. However, if we were to do this again, we would invite both the women scientists and parents to attend a culminating showcase event to bring more of the unfamiliar and familiar world beyond school in to the classroom. This kind of exchange would provide important eyes and ears to witness and to hear the students and support their learning. We had planned to hold a culminating event like this; however, we ran out of time and the culminating event included the girls, teachers, and researchers. It would be wonderful to have the women in science and parents witness the remarkable work of these young adults.

Creating Pathways via Literacy—A Call to Action

In this book, we offer a model of how teachers, literacy coaches, writing project sites, and university scholars may create literacy communities within or beyond their classrooms as a means of creating more equitable learning for Latina girls to imagine, investigate, and articulate their future selves. This study serves as an example of how literacy researchers and teachers may examine the teaching and learning of reading and writing as acts deeply grounded in the sociocultural activities, curricular invitations, and available sponsors within and beyond school settings (Brandt, 2001; Early, 2010; Prior, 2006). The structure and content GWS created the conditions for the social construction of a new kind of literacy learning the girls used to envision their possible selves in science (Oyserman et al., 2002). This project also serves as a poignant model for researchers and educators committed to working toward equity and inclusion in education, shrinking the gender divide in the sciences, and providing more equitable opportunities for adolescent girls across grade levels and disciplines.

GWS serves as an example of how the research and teaching of literacy may offer new learning pathways for Latina girls, and other adolescent girls, to connect to the larger world beyond their lived experiences and in relation to others to think about how they want to act, live, and work in the world. This study also suggests potential opportunities for longitudinal research beginning with real-world writing experiences within school settings to document its impact and influence on women in college, the workplace, and the community. This curricular model has the potential to cut across disciplines, beyond science, and for various purposes. There are many obstacles to overcome if we are to raise the number of women, especially women of color, involved in scientific careers and pathways. This project suggests that part of the solution occurs in the creation of literacy communities that value reading and writing, so students have opportunities to access, engage with, and write for real audiences as a way to envision possibilities for their future selves. As we close this book, we encourage you to question: What would this work look like in your school? Research site? After-school program? Community center? Imagine the possibilities and begin there.

A Call to Action

Design Thinking Process Guide

Taking the concept of designing a literacy community of practice and making it your own. Here are helpful questions to help you start thinking about ways you can tell the model will share in this book and design your own.

What is the project?
Who is the audience/user for this project?
What problem does it try to solve?
What is the purpose of this project?
How will this project building sponsorship or support for the participants?
Where will the project take place?
When?
How?
What are your inquiry questions?

Bibliography

Altieri, J. (2016). *Reading science: Practical strategies for integrating instruction.* Portsmouth, NH: Heinemann.

Applebee, A. N., & Langer, J. A. (2011). A snapshot of writing instruction in middle schools and high schools. *English Journal, 100*(6), 14–27.

Applebee, A. N., & Langer, J. A. (2013). *Writing instruction that works. Proven methods for middle and high school classrooms*. New York, NY: Teachers College Press.

Applebee, A. N., Langer, J. A., & Mullis, I. V. S. (1974/1986). *Writing: Trends across the decade, 1974–84. National Assessment of Educational Progress*. Princeton, NJ: Educational Testing Service.

Ashcraft, C., Eger, E., & Scott, K. A. (2017). A tale of two cohorts: Engaging a diverse range of girls in technology through culturally responsive computing. *Anthropology and Education Quarterly, 48*(3), 233–251.

Atwell, N. (1990). *Coming to know: Writing to learn in the intermediate grades.* Portsmouth, NH: Heinemann.

Babcock, L., & Laschever, S. (2007). *Women don't ask: Negotiation and the gender divide*. Princeton, NJ: Princeton University Press.

Baker, T., & Carroll, B. (2016). Working at the intersections of formal and informal science and literacy education. *Connected Science Learning*, (1), Retrieved June 17, 2017, from http://csl.nsta.org/2016/03/working-at-the-intersections/

Bakhtin, M. M. (2010). *Speech genres and other late essays*. Austin: University of Texas Press.

Bandura, A. (1986). *Social foundations of thought and action: A social cognitive theory*. Englewood Cliffs, NJ: Prentice-Hall.

Bangert-Drowns, R. L., Hurley, M. M., & Wilkinson, B. (2004). The effects of school-based writing-to-learn interventions on academic achievement: A meta-analysis. *Review of Educational Research, 74*(1), 29–58.

Bazerman, C. (2010). 2009 CCCC Chair's address: The wonder of writing. *College Composition and Communication, 61*(3), 571–580.

Bazerman, C. (2016). What do sociocultural studies of writing tell us about learning to write? chapter 1. In C. A. MacArthur, S. Graham, & J. Fitzgerald (Eds.), *Handbook of writing research* (pp. 11–23). New York, NY: Guilford Press.

Bazerman, C., Applebee, A., Berninger, V. W., Brandt, D., Graham, S., Matsuda, P. K., . . . Schleppegrell, M. (2017). Taking the long view on writing development. *Research in the Teaching of English, 51*(3), 351–360.

Bazerman, C., Bonini, A., & Figueiredo, D. (Eds.). (2009). *Genre in a changing world*. Fort Collins, CO: Parlor Press and the WAC Clearinghouse.
Bean, J. C., Drenk, D., & Lee, F. D. (1982). Microtheme strategies for developing cognitive skills. In C. W. Griffin (Ed.), *Teaching writing in all disciplines* (pp. 27–38). San Francisco, CA: Jossey-Bass.
Bereiter, C., & Scardamalia, M. (1987). *The psychology of written composition*. Mahwah, NJ: Lawrence Erlbaum Associates.
Bian, L., Leah, S. J., & Cimpian, A. (2017). Gender stereotypes about intellectual ability emerge early and influence children's interests. *Science*, *355*(6323), 389–391.
Bielby, R., Posselt, J. R., Jaquette, O., & Bastedo, M. N. (2014). Why are women underrepresented in elite colleges and universities? A non-linear decomposition analysis. *Research in Higher Education*, *55*(8), 735–760.
Brandt, D. (2001). *Literacy in American lives*. Cambridge: Cambridge University Press.
Brenner, M. E. (2006). *Interviewing in educational research. Handbook of complementary methods in education research* (2nd ed.). New York, NY: Routledge, pp. 357–370.
Bridwell, L. S. (1980). Revising strategies in twelfth grade students' transactional writing. *Research in the Teaching of English*, *14*, 197–222.
Britton, J. (1970/1972). *Language and learning*. Portsmouth NH: Boynton/Cook, Heinemann.
Burke, R. J., & Mattis, M. C. (Eds.). (2007). *Women and minorities in science, technology, engineering, and mathematics: Upping the numbers*. Northhampton, MA: Edward Elgar Publishing.
CCSS (Common Core State Standards Initiative). (2011). *Common core state standards for English language arts & literacy in history/social studies, science, and technical subjects*. Washington, DC: National Governors Association Center for Best Practices and the Council of Chief State School Officers. Retrieved June 16, 2017, from www.corestandards.org/read-the-standards/
Chambers, D. W. (1983). Stereotypic images of the scientist: The draw-a-scientist test. *Science Education*, *67*(2), 255–265.
Christensen, L. (2000). *Reading, writing, and rising up: Teaching about social justice and the power of the written word*. Milwaukee, WI: Rethinking Schools.
Colucci, L. (2014). *Why we should all hack medicine*. Brussels: TEDx. Retrieved June 26, 2018, from www.tedxbrussels.eu/lina-colucci/
Compton-Lilly, C. (2009). The development of habitus over time. WCER Working Paper No. 2009–7. Madison: University of Wisconsin—Madison, Wisconsin Center for Education Research. Retrieved June 13, 2018, from https://eric.ed.gov/?id=ED506375
Copeland, M. (2005). *Socratic circles: Fostering critical and creative thinking in middle school and high school*. Portland, ME: Stenhouse Publishers.
Cox, M. D. (2004). Introduction to faculty learning communities. *New Directions for Teaching and Learning*, (97), 5–23.
Crenshaw, K. (1989). Demarginalizing the intersection of race and sex: A black feminist critique of antidiscrimination doctrine, feminist theory and antiracist politics. *University of Chicago Legal Forum*, 139.

Crenshaw, K. (1991). Mapping the margins: Intersectionality, identity politics, and violence against women of color. *Stanford Law Review*, 1241–1299.

Dean, D. (2008). *Genre theory: Teaching, writing, and being*. Urbana, IL: National Council of Teachers of English.

deVos, A. (2014). Why you should care about whale poo. Rio De Janeiro, Brazil: TEDGlobal. Retrieved June 26, 2018, from www.ted.com/talks/asha_de_vos_why_you_should_care_about_whale_poo#t-11683

Diekman, A. B., Brown, E. R., Johnston, A. M., & Clark, E. K. (2010). Seeking congruity between goals and roles: A new look at why women opt out of science, technology, engineering, and mathematics careers. *Psychological Science*, *21*(8), 1051–1057.

Duke University Office of Undergraduate Scholars & Fellows. (Fall, 2014). Designing cool stuff. *Distinction the Scholars Magazine*. Retrieved from https://issuu.com/cholcomb/docs/distinction_110514_final_962bf965a2f6e0

Dweck, C. S. (2006). *Mindset: The new psychology of success*. New York: Random House.

Dyson, A. H. (2003). Welcome to the Jam: Popular Culture, School Literacy, and the Making of Childhoods. *Harvard Educational Review*, *73*(3), 328–361.

Eagleton, M., Guinee, K., & Langlais, K. (2003). Teaching internet literacy strategies: The hero inquiry project. *Voices from the Middle*, *10*(3), 28.

Early, J. S. (2006). *Stirring up justice: Writing and reading to change the world*. Portsmouth, NH: Heinemann.

Early, J. S. (2010). 'Mi Hija, You Should Be a Writer': The role of parental support and learning to write. *Bilingual Research Journal*, *33*(3), 277–291.

Early, J. S. (2017). This is who I want to be!: Exploring possible selves by interviewing women in science. *Journal of Adolescent & Adult Literacy*, *61*(1), 75–83.

Early, J. S., & DeCosta, M. (2012). *Real world writing for secondary students: Teaching the college admission essay and other gate-openers for higher education*. New York, NY: Teachers College Press.

Early, J. S., & DeCosta-Smith, M. (2011). Making a case for college: A genre-based college admission essay intervention for underserved high school students. *Journal of Writing Research*, *2*(3), 299–329.

Early, J. S., & Flores, T. (2017). Escribiendo juntos: Toward a collaborative model of multiliterate family literacy in English-only and anti-immigrant contexts. *Research in the Teaching of English*, *52*(2), 156–180.

Early, J. S., & Saidy, C. (2014a). A study of a multiple component feedback approach to substantive revision for secondary ELL and multilingual writers. *Reading and Writing*, *27*(6), 995–1014.

Early, J. S., & Saidy, C. (2014b). Uncovering substance. *Journal of Adolescent & Adult Literacy*, *58*(3), 209–218.

Elliott, L. A., Jaxon, K., & Salter, I. (2016). *Composing science: A facilitator's guide to writing in the science classroom*. New York, NY: Teachers College Press.

Erickson, F. (1985). *Qualitative methods in research on teaching*. Occasional Paper No. 81. Retrieved June 29, 2017, from http://eric.ed.gov/?id=ED263203

Espinosa, L. (2011). Pipelines and pathways: Women of color in undergraduate STEM majors and the college experiences that contribute to persistence. *Harvard Educational Review, 81*(2), 209–241.

Finson, K. D. (2002). Drawing a scientist: What we do and do not know after fifty years of drawings. *School Science and Mathematics, 102*(7), 335–345.

Fishman, J., Lunsford, A., McGregor, B., & Otuteye, M. (2005). Performing writing, performing literacy. *College Composition and Communication*, 224–252.

Fleischer, C., & Andrew-Vaughan, S. (2009). *Writing outside your comfort zone: Helping students navigate unfamiliar genres*. Portsmouth, NH: Heinemann.

Frankel, L. P. (2014). *Nice girls don't get the corner office: Unconscious mistakes women make that sabotage their careers*. New York, NY: Business Plus.

Freedman, M. P. (2002). The influence of laboratory instruction on science achievement and attitude toward science across gender differences. *Journal of Women and Minorities in Science and Engineering, 8*(2), 191–200.

Gallagher, K. (2011). *Write like this: Teaching real-world writing through modeling & mentor texts*. Portland, ME: Stenhouse Publishers.

Gawande, A. (2007). *Better: A surgeon's notes on performance*. New York, NY: Picador Press.

Graham, S., & Harris, K. H. (2005). *Writing better: Effective strategies for teaching students with learning disabilities*. Baltimore, MA: Brookes Publishing.

Graham, S., & Perin, D. (2007). *Writing next: Effective strategies to improve writing of adolescents in middle and high schools*. A Report to Carnegie Corporation of New York, Alliance for Excellent Education.

Guarino, C. M., Santibañez, L., & Daley, G. A. (2006). Teacher recruitment and retention: A review of the recent empirical literature. *Review of Educational Research, 76*(2), 173–208.

Gunel, M., Hand, B., & Prain, V. (2007). Writing for learning in science: A secondary analysis of six studies. *International Journal of Science and Mathematics Education, 5*(4), 615–637.

Harklau, L., & Pinnow, R. (2009). Adolescent second-language writing. In L. Christenbury, R. Bomer, & P. Smagorinsky (Eds.) *Handbook of adolescent literacy research* (pp. 126–139). New York, NY: Guilford Press.

Harris, J. (2006). *Rewriting: How to do things with texts*. Logan, UT: Utah State UP.

Harwell, S. (2000). In their own voices. Middle level girls' perceptions of teaching and learning science. *Journal of Science Teacher Education, 11*(3), 221–242.

Heard, G. (1999). *Awakening the heart: Exploring poetry in elementary and middle school*. Portsmouth, NH: Heinemann.

Heard, G. (2016). *Heart maps: Helping students create and craft authentic writing*. Portsmouth, NH: Heinemann.

Hicks, T., & Turner, K. H. (2013). No longer a luxury: Digital literacy can't wait. *English Journal*, 58–65.

Hill, C., Corbett, C., & St. Rose, A. (2010). *Why so few?: Women in science, technology, engineering and mathematics*. Washington, DC: American Association of University Women. Retrieved June 28, 2017, from www.aauw.org/

files/2013/02/Why-So-Few-Women-in-Science-Technology-Engineering-and-Mathematics.pdf

Kittle, P. (2008). *Write beside them: Risk, voice, and clarity in high school writing*. Portsmouth, NH: Heinemann.

Klein, P. D. (2000). Elementary students' strategies for writing-to-learn in science. *Cognition and Instruction, 18*(3), 317–348.

Langer, J. A., & Applebee, A. N. (1987). *How writing shapes thinking: A study of teaching and learning*. NCTE Research Report No. 22. Urbana, IL: National Council of Teachers of English.

Lave, J., & Wenger, E. (1999). Learning and pedagogy in communities of practice. *Learners and Pedagogy*, 21–33.

Lee, J., Husman, J., Scott, K. A., & Eggum-Wilkens, N. D. (2015). Compugirls: Stepping stone to future computer-based technology pathways. *Journal of Educational Computing Research, 52*(2), 199–223.

Maranto, R., Milliman, S., Hess, F., & Gresham, A. (2008). *School choice in the real world: Lessons from Arizona charter schools*. Boulder, CO: Westview Press.

McIntosh, M. E., & Draper, R. J. (2001). Using learning logs in mathematics: Writing to learn. *The Mathematics Teacher, 94*(7), 554–557.

McLeod, S. H., & Soven, M. (1992). *Writing across the curriculum*. Newbury Park, CA: Sage.

Miller, P. H., Slawinski Blessing, J., & Schwartz, S. (2006). Gender differences in high-school students' views about science. *International Journal of Science Education, 28*(4), 363–381.

Mitra, D. L. (2008). Balancing power in communities of practice: An examination of increasing student voice through school-based youth—adult partnerships. *Journal of Educational Change, 9*(3), 221.

Moll, L. C., Amanti, C., Neff, D., & González, N. (1992). Funds of knowledge for teaching: Using a qualitative approach to connect homes and classrooms. *Theory into Practice, 31*, 132–141.

National Academy of Sciences. (2012). *A Framework for K-12 Science Education*. Retrieved September 7, 2018 from http://nextgenscience.org/sites/default/files/Final%20Release%20NGSS%20Front%20Matter%20-%206.17.13%20Update_0.pdf

National Center for Education Statistics. (2013–2014). *Search for public schools data*. Retrieved April 14, 2017, from http://nces.ed.gov/ccd/schoolsearch/index.asp

National Research Council. (2012). *A framework for K-12 science education: Practices, crosscutting concepts, and core ideas*. Washington, DC: The National Academies Press.

National Science Foundation. (2012). *Building informal science education and literacy partnerships: A collaborative project of the National Writing Project and the Association of Science-Technology Centers*. Award Number: 1224161. Retrieved on June 15, 2017, from www.nsf.gov/awardsearch/showAward?AWD_ID=1224161&HistoricalAwards=false

Nevarez, G., & Wyloge, E. (2016). *Arizona school data shows uneven distribution of ethnic groups*. Arizona Center for Investigative Reporting. Retrieved June 25, 2018, from https://azcir.org/news/2016/02/12/arizona-school-ethnicity-disparity-charter-district/

Newell, G. E. (2006). Writing to learn: How alternative theories of school writing account for student performance. In C. A. MacArthur, S. Graham, & J. Fitzgerald (Eds.), *Handbook of writing research* (pp. 235–247). New York, NY: Guilford Press.
Olson, C. B., & Land, R. (2007). A cognitive strategies approach to reading and writing instruction for English language learners in secondary school. *Research in the Teaching of English*, 269–303.
Ong, M., Wright, C., Espinosa, L., & Orfield, G. (2011). Inside the double bind: A synthesis of empirical research on undergraduate and graduate women of color in science, technology, engineering, and mathematics. *Harvard Educational Review*, *81*(2), 172–209.
Oyserman, D., Terry, K., & Bybee, D. (2002). A possible selves intervention to enhance school involvement. *Journal of Adolescence*, *25*(3), 313–326.
Pain, E. (2015). En pointe. *Science Magazine*, *348*, 366.
Parker, L. H., & Rennie, L. J. (2002). Teachers' implementation of gender-inclusive instructional strategies in single-sex and mixed-sex science classrooms. *International Journal of Science Education*, *24*(9), 881–897.
Prior, P. (2006). A sociocultural theory of writing. In C. MacArthur, S. Graham, & J. Fitzgerald (Eds.), *The handbook of writing research* (pp. 54–66). New York, NY: Guilford Press.
Purcell-Gates, V., Duke, N. K., & Martineau, J. A. (2007). Learning to read and write genre-specific text: Roles of authentic experience and explicit teaching. *Reading Research Quarterly*, *42*(1), 8–45.
Raphael, T. E. (1986). Teaching question answer relationships, revisited. *The Reading Teacher*, *39*(6), 516–522.
Richardson, A. E. (2012). *Explainers' development of science-learner identities through participation in a community of practice*. Keene, NH: Antioch University.
Ryan, K. J., Myers, N., & Jones, R. (2016). Introduction: Identifying feminist ecological ethe. In K. J. Ryan, N. Myers, & R. Jones (Eds.), *Rethinking ethos: A feminist ecological approach to rhetoric* (pp. 1–22). Carbondale, IL: SIU Press.
Saidy, C. (2013). Working from the inside out: Writing for community and democratic participation when citizenship is in question. *English Journal*, 60–65.
Saidy, C. (2017). Girls writing science: Opening up access in a girls' reading and writing group. *English Journal*, *106*(5), 27.
Saidy, C., & Early, J. S. (2016). You need more organization bro: Relationship building in secondary writing and revision. *The Clearing House: A Journal of Educational Strategies, Issues and Ideas*, *89*(2), 54–60.
Saul, W., Reardon, J., Pearce, C. R., Dieckman, D., & Neutze, D. (2002). *Science workshop: Reading, writing, and thinking like a scientist* (2nd ed.). Portsmouth, NH: Heinemann.
Scantlebury, K., Baker, D., Suga, Yoshida, A., & Uysal, S. (2007). Avoiding the issue of gender in Japanese science education. *International Journal of Science and Mathematics Education*, *5*(3), 415–438.
Scott, K. A., & Clark, K. (2013). Digital engagement for urban youth: From theory to practice. *Urban Education*, *48*(5), 627–628.
Scott, K. A., & White, M. A. (2013). COMPUGIRLS' standpoint: Culturally responsive computing and its effect on girls of color. *Urban Education*, *48*(5), 657–681.

Shell, D. F., Colvin, C., & Bruning, R. H. (1995). Self-efficacy, attributions, and outcome expectancy mechanisms in reading and writing achievement: Grade-level and achievement level differences. *Journal of Educational Psychology*, *87*, 386–398.

Siann, G., & Callaghan, M. (2001). Choices and barriers: Factors influencing women's choice of higher education in science, engineering and technology. *Journal of Further and Higher Education*, *25*(1), 85–95.

Sim, C. (2006). Preparing for professional experiences—incorporating pre-service teachers as 'communities of practice'. *Teacher and Teacher Education*, *22*(1), 77–83.

Simard, Suzanne. (2016). *How trees talk to each other*. TED talk. Banff, Alberta: TED Summit. Retrieved June, 2018, from www.ted.com/talks/suzanne_simard_how_trees_talk_to_each_other#t-7033

Singer, J., & Hubbard, R. (2003). Teaching from the heart: Guiding adolescent writers to literate lives. *Journal of Adolescent & Adult Literacy*, *46*, 326–338.

Skloot, R. (2011). *The immortal life of Henrietta Lacks*. New York, NY: Broadway Books.

Spradley, J. P. (2016). *Participant observation*. Long Grove, IL: Waveland Press.

Swales, J. M. (1988). Discourse communities, genres and English as an international language. *World Englishes*, *7*(2), 211–220.

US Census Data. (2017). *Public education finances: 2015*. Washington, DC: US Government Printing Office. Retrieved June 16, 2018, from www.census.gov/content/dam/Census/library/publications/2017/econ/g15-aspef.pdf

Vescio, V., Ross, D., & Adams, A. (2008). A review of research on the impact of professional learning communities on teaching practice and student learning. *Teaching and Teacher Education*, *24*(1), 80–91.

Volosinov, V. (1973). *Marxism and the philosophy of language*. (L. Matejka & I. Titunik, Trans.). Cambridge, MA: Harvard University Press.

Vygotsky, L. S. (1978). *Mind in society: The development of higher psychological processes*. Cambridge, MA: Harvard University Press.

Wardle, E., & Roozen, K. (2012). Addressing the complexity of writing development: Toward an ecological model of assessment. *Assessing Writing*, *17*(2), 106–119.

Wenger, E. (1998). *Communities of practice: Learning, meaning, and identity*. Cambridge: Cambridge University Press.

Wenger, E. (2006). *Communities of practice: A brief introduction*. Retrieved May 28, 2017, from www.ewenger. com/theory/communities_of_practice_intro. htm (2008).

Wenger, E., McDermott, R. A., & Snyder, W. (2002). *Cultivating communities of practice: A guide to managing knowledge*. Boston, MA: Harvard Business Press.

White, W. (1989). Her deepness. Retrieved May 2, 2018, from *The New Yorker*. www.newyorker.com/magazine/1989/07/03/deepness

Whitney, A. (2008). Teacher transformation in the National Writing Project. *Research in the Teaching of English*, *43*(2), 144–187.

Wiley, M. (2000). The popularity of formulaic writing (and why we need to resist). *The English Journal*, *90*(1), 61–67.

Winn, M. (2011). *Girl time: Literacy, justice, and the school-to-prison pipeline*. New York, NY: Teachers College Press.

Yosso, T. J. (2005). Whose culture has capital? A critical race theory discussion of community cultural wealth. *Race Ethnicity and Education*, *8*(1), 69–91.

Women in Science Bibliography

Bailey, E. T. (2016). *The sound of a wild snail eating*. New York, NY: Algonquin Books.
Baker, R. (1944). *The first woman doctor: The story of Elizabeth Blackwell, MD*. New York: Julian Messner.
Beaty, A. (2013). *Rosie Revere, engineer*. New York, NY: Abrams.
Beyer, K. W. (2015). *Grace Hopper and the invention of the information age*. Pennsauken, NJ BookBaby.
Breen, S. (2016). *Violet the pilot*. New York: Puffin Books.
Brighton, C. (2000). *The fossil girl: Mary Anning's dinosaur discovery*. Brookfield, CT: Lincoln.
Burleigh, R. (2013). *Look Up!: Henrietta Leavitt, Pioneering Woman Astronomer*. New York: Simon & Schuster.
Chin-Lee, C. (2008). *Amelia to Zora: Twenty-six women who changed the world*. Watertown, MA: Charlesbridge.
Clay, L. (2009). *Summer birds: The butterflies of Maria Merian*. New York: Henry Holt and Company.
Deakin, M. A. (2007). *Hypatia of Alexandra: mathematician and martyr (astronomer as well)*. Amherst, NY: Prometheus Books.
Edwards, R., & O'Brien, J. (2012). *Who is Jane Goodall*? New York, NY: Grosset & Dunlap.
Emling, S. (2009). *The fossil hunter*. New York, NY: Palgrave Macmillan.
Essinger, J. (2014). *Ada's algorithm: How Lord Byron's daughter Ada Lovelace launched the digital age*. New York, NY: Melville House.
Goodall, J. (2010). *Through a window: Thirty years with the chimpanzees of Gombe*. Boston, MA: Houghton Mifflin Harcourt.
Harris, A., & Rowe, J. A. (2013). *I wonder*. Boston, MA: Houghton Mifflin Harcourt/ Four Elephants Press.
Hitchcock, S. T. (2004). *Rita Levi-Montalcini: Nobel prize winner (Women in medicine)*. Langhorne, PA: Chelsea House Publishers.
Ignotofsky, Rachel. (2016). *Women in science: 50 fearless pioneers who changed the world*. New York, NY: Ten Speed Press.
Kelly, J. (2009). *The evolution of calpurnia tate* (Vol. 1). New York, NY: Henry Holt & Company.
Kiernan, D. (2014). *The girls of atomic city: The untold story of the women who helped win World War II*. New York, NY: Simon & Schuster.
Lester, A. (2013). *Sophie Scott goes south*. Boston, MA: Houghton Mifflin Harcourt.
Maddox, B. (2002). *Rosalind Franklin: The dark lady of DNA*. New York, NY: HarperCollins.
McCully, E. A. (2013). *Marvelous Mattie: How Margaret E. Knight became an inventor*. New York: Farrar, Straus and Giroux.
McDonnell, P. (2011). *Me . . . Jane*. Boston, MA: Little, Brown.
McGrayne, S. B. (1998). *Nobel Prize women in science: Their lives, struggles, and momentous discoveries*. Secaucus, NJ: Carol Pub. Group.
Meltzer, B., & Eliopoulos, C. (2016). *I am Jane Goodall*. New York, NY: Penguin Books.

Montgomery, S. (2012). *Temple Grandin: How the girl who loved cows embraced autism and changed the world.* Boston, MA: Houghton Mifflin Harcourt.
Morgan, G. D., & Stroupe, A. (2017). *Rocket girl: The story of Mary Sherman Morgan, America's first female rocket scientist.* Solon, OH: Findaway World LLC.
Peterson, D. (2014). *Jane Goodall: The woman who redefined man.* Boston, MA: Houghton Mifflin Harcourt.
Sime, R. L. (1996). *Lise Meitner: A life in physics* (Vol. 11). Berkeley, CA: University of California Press.
Skloot, R., & Turpin, B. (2010). *The immortal life of Henrietta Lacks.* New York, NY: Broadway Books.
Spangenburg, R., & Moser, D. K. (2008). *Barbara McClintock: Pioneering geneticist (Makers of modern science).* Langhorne, PA: Chelsea House Publishers.
Spires, A. (2014). *The most magnificent thing.* Toronto, Canada: Kids Can Press Ltd.
Stone, T. L. (2017). *Who says women can't be doctors?* Danbury, CT: Woods Studios Inc.
Straus, E. (2000). *Rosalyn Yalow: Nobel laureate: Her life and work in medicine.* New York: Perseus Books.
Swaby, R. (2015). *Headstrong: 52 women who changed science-and the world.* New York, NY: Broadway Books.
Todd, K. (2007). *Chrysalis: Maria Sibylla Merian and the secrets of metamorphosis.* Boston: MA: Houghton Mifflin Harcourt.
Winter, J. (2011). *The watcher: Jane Goodall's life with chimps.* Toronto, Ontario: Schwartz & Wade.

Index

Note: Page numbers in *italics* indicate figures, and page numbers in **bold** indicate tables on the corresponding page.

For Product Safety Concerns and Information please contact our EU
representative GPSR@taylorandfrancis.com
Taylor & Francis Verlag GmbH, Kaufingerstraße 24, 80331 München, Germany

www.ingramcontent.com/pod-product-compliance
Ingram Content Group UK Ltd.
Pitfield, Milton Keynes, MK11 3LW, UK
UKHW022157190926
475702UK00003B/78

* 9 7 8 0 3 6 7 0 0 1 0 8 7 *